Praise for *Wayfinding*

"This book is a quick read that delivers clear, practical insights that leaders can immediately apply in their day-to-day decisions. The story and frameworks deliver concrete, actionable messages that help teams move from ideas to measurable impact."

– Steve Rohleder, Chair of the Board, Cognizant, and former Global COO, Accenture

"Thank you to Stephen and Ben for taking a confusing topic and making it easy to understand. I run a midsize, tech-based services company and the challenges, discussions, and solutions captured in this book are real-life representations of what is going on in our company today. Most CEOs need practical guidance and approaches when it comes to figuring out how to effectively incorporate AI into our overall business strategy, and *Wayfinding* is an excellent resource for any leadership team that wants to be better educated and enlightened about AI."

– J. Heffy Provost, CEO, AGDATA

"I wish every CEO facing the 'AI question' had this book in hand. It's the no-nonsense guide we all need right now."

– Alessia Falsarone, Independent Director, Generali, and Adjunct Faculty, University of Chicago

"AI begets hyperbole. Heroes and Villains. AI books too: They're typically either breathless or brittle. In *Wayfinding*, Stephen Straus and Ben Szuhaj avoid these cartoonish corners and, through vivid stories and clear frameworks, help us move from anxiety and theatrics to practical tactics."

– Mike Bechtel, former Chief Futurist, Deloitte, and Adjunct Professor, University of Notre Dame

"In *Wayfinding*, Stephen Straus and Ben Szuhaj tackle the single biggest challenge facing founders today: how to lead when the rules are being rewritten in real time. Through a compelling business fable, Straus and Szuhaj show what it actually takes to navigate AI-driven disruption—not with theory, but with hard-earned lessons from the field. If you're building a firm in the age of AI, this book will sharpen your thinking and accelerate your path forward."

– Greg Alexander, Founder and CEO, Collective 54

"In a moment when AI conversations generate more heat than light, *Wayfinding* delivers something rare: genuine clarity. Straus and Szuhaj use the power of narrative to cut through complexity, offering leaders a coherent framework for thinking about AI that respects both the technology's potential and its limits. This is the book I wish more AI advocates would read."

– Stephen Klein, Founder and CEO, Curiouser.AI, and AI Lecturer, UC Berkeley

"*Wayfinding* isn't just a timely guide—it's an urgent call for leaders navigating the world of AI. Through engaging storytelling and practical wisdom, Stephen and Ben show how to build the systems and teams that win in a rapidly changing landscape. Every CEO and business builder should read this book if they want to turn disruption into sustainable growth."

– John Arnott, CEO, C1M.ai

Wayfinding

Wayfinding

A Business Fable About Leading in the Age of AI

Stephen Straus and Ben Szuhaj

Matt Holt Books
An Imprint of BenBella Books, Inc.
Dallas, TX

MATT HOLT BENBELLA

Matt Holt is an imprint of BenBella Books, Inc.
BenBella Books, Inc.
8080 N. Central Expressway
Suite 1700
Dallas, TX 75206
benbellabooks.com
Send feedback to feedback@benbellabooks.com

Matt Holt and *BenBella* are federally registered trademarks.

Printed in the United States of America
10 9 8 7 6 5 4 3 2 1

Library of Congress Control Number: 2026018740
ISBN 9781637749845 (hardcover)
ISBN 9781637749852 (electronic)

Editing by Lydia Choi
Copyediting by Michael Fedison
Proofreading by Denise Pangia and Becky Maines
Text design and composition by Jordan Koluch
Cover design by Jason Arias
Cover Image © Adobe Stock / Brocreative
Printed by Lake Book Manufacturing

This book is dedicated to the team at KUNGFU.AI
who are helping our clients create transformative value
through innovative and responsible AI solutions every day
with the goal of helping build a world that harnesses AI
for the unprecedented benefit of people and the planet.

Contents

Act 3

Preface: AI or Goodbye

Typically, leaders are at the top of their game, with the expertise and experience needed to succeed. But the largest technological transition of any of our lifetimes has shifted the whole playing field and fundamentally changed the basis of competition. This has left leaders in the uncomfortable position of having neither the expertise nor the experience needed to navigate this new era.

Leaders across the board are all facing this unprecedented challenge, and the stakes couldn't be higher, both for their careers and their organizations.

But where to turn for good, practical advice? There's a lot of hype out there, and it's easy to get distracted and focus on things that won't drive the kind of substantial and differentiated value it takes to really succeed. While there are endless numbers of new books on AI, most of these authors are new to the technology since ChatGPT launched at the end of 2022.

The original frameworks and learnings in *Wayfinding* are hard-won from eight years of in-the-trenches experience of an

AI-native professional services firm developing leading-edge strategies and bespoke AI solutions on the cutting edge of the technology for organizations large and small across a wide range of industries.

We love business fables—they're quick and easy to read, and we find them to be a great way to learn. While all of the characters and the companies they work for in this book are fictitious, many aspects of the book are composites of people and companies we've worked with, the situations they've found themselves in, and the decisions they've made. And we end the book with an appendix summarizing all the methodologies and key learnings, as well as links to more details, a glossary, and FAQs, all of which can be found at wayfindingbook.com.

Welcome to *Wayfinding*, a proprietary methodology for managing AI and a metaphor for leading in this new age of AI. We hope that you find this book to be practical, timely, and timeless as well as bingeable. It's for anyone looking to up their AI game, whether you're just getting started or are far along in your journey.

Character List

Main Characters

- **Sarah Foster:** CEO of FreightCo Shipping, narrator
- **Julie Guthrie:** President of FreightCo's South Central Division
- **Mark Armstrong:** President of FreightCo's Western Division
- **Paul Walcott:** FreightCo board member, Sarah's mentor

Sarah's Family

- **David Brown:** Husband, policy consultant, and professor
- **Emily Brown:** Daughter, 16
- **Jake Brown:** Son, 13
- **Lily Brown:** Daughter, 11

FreightCo Board

- **Richard:** Board chair
- **Diane:** Board member
- **Michael:** Board member
- **James:** Board member

Austin Division Team

- **Ross:** CTO
- **August:** Head of People
- **Raj:** Operations manager
- **Anita:** Contracting team member
- **Andy:** Machine learning engineer

Consulting Partners

- **Oliver:** Cofounder and CEO of Guidebridge.ai
- **Ava:** Head of AI Strategy at Guidebridge.ai

Other CEOs

- **Noah:** CEO of Greentown Pharmacy
- **Dante:** CEO of a SaaS company
- **Maya:** CEO from financial services
- **Phillip:** CEO of Shop & Buy (S&B), major FreightCo client

Other Companies

- **FreightCo Shipping:** 35-year-old logistics company
- **Apex Shipping:** FreightCo's main competitor
- **Shop & Buy (S&B):** FreightCo's largest client
- **Guidebridge.ai:** AI consulting firm that partners with Julie's division
- **Weyco Industries:** FreightCo client
- **Care Alliance:** FreightCo client

Introduction

The Ultimatum

The boardroom fell silent. I could tell something big was coming by the way Richard leaned forward in his chair to address me.

"I'll be blunt, Sarah," he finally said. "You have until next quarter to show us a real AI strategy. Otherwise, we'll need to make a leadership change."

I gripped my pen tightly, fighting to keep my face neutral. The pen had been a gift from Paul Walcott—my mentor, who sat across the table from me now—when I became CEO. "For signing the big decisions," he'd said with a wink.

It dawned on me now with a mix of shock and bemused detachment that for the past half hour I had been using it to jot down notes about my potential firing. I'd walked into this room expecting appreciation for our surging stock price and the best quarterly results in company history. Instead, I felt I was being blindsided.

"Companies that don't embrace AI will be also-rans in a few years." Richard tapped his notepad for emphasis. "The board believes—and I agree—that without a comprehensive AI strategy, FreightCo Shipping risks becoming irrelevant, regardless of our current performance. This is something we've talked about in multiple board meetings before. But we haven't seen real concrete action from you on AI. So consider this an ultimatum."

I wanted to protest that I hadn't been ignoring AI entirely. I *had* been sharing some of my thoughts about it with the board. But how could I focus fully on it when there were more immediate problems to solve? And was it really so goddamn urgent and important?

I glanced around the table. Eight faces, all watching me. Some sympathetic, some impatient. No one spoke up in my defense. James checked his phone under the table. Diane tapped her smart watch, her expression unreadable. Michael, normally a reliable ally, studied the table. Paul caught my gaze briefly, adjusting his glasses. I could tell he was agitated.

"Our largest client, Shop & Buy, explicitly asked about our AI capabilities last week," Richard added. "That's $200 million in annual revenue questioning whether we can meet their future needs. You need to make sure we don't lose them as a client. And, as you know, our competitors are announcing AI investments almost weekly. The few we've made have helped relieve investor scrutiny for now, but in this room, I think we all know that we're not leading the market in AI. We're not even keeping up with the market. We've had a chance to discuss this, and we believe what's needed is a clear AI road map for the company, one that builds on our unique strengths as an organization and positions us as a leader in this new age of AI over the next three years. Is that clear?"

"Yes," I said. "Clearly this has been a closed-door topic of conversation. Have you discussed what our AI budget should look like?"

"Good question. We'll talk about that once we review what you've put together at our next quarterly meeting."

Richard closed his leather portfolio with finality. Just like that, it was over.

I walked out of the boardroom.

In the empty hallway, I paused to steady myself. FreightCo was thriving by every metric. Our logistics management platform had been industry-leading for years, helping companies track shipments and manage inventory with precision. Last quarter's numbers were the best in company history. Our market share was growing. We'd won three industry awards this year. And yet, I was sure they were going to fire me.

For just a moment, I let my mask slip—eyes closed, I swallowed against the tightness in my throat.

I made my way to my office. I tried to focus on the small and achievable tasks in front of me, but my mind kept wandering. The AI initiatives I'd approved—the chatbot for customer service, the predictive maintenance pilot—suddenly seemed woefully inadequate. I got up and stood by the floor-to-ceiling window. Midtown Atlanta looked monstrous.

My eyes shifted focus, and I noticed my reflection: dark hair in a practical bob, tailored navy suit. I cut the composed figure of someone who was supposed to have all the answers. But I felt like a child again.

In my head I was still that sixteen-year-old girl stocking shelves at the grocery store in Pittsburgh, waiting for Mom's nursing shift to end. I was back in college, folding clothes on weekend shifts while my classmates partied. I was that anxious entry-level associate who read and reread every email five times before sending. Twenty years of work compressed into a flash of memory. Marketing coordinator. Brand strategist. CMO. And now CEO. All that climbing, all that proving, and in this moment it felt like none of it mattered. I had ninety days to master a technology that had barely existed a few years ago—a technology I barely understood—and that, as far as I could tell, was mostly being used by students to cheat on their homework. I frowned at the window, anger rising at the unfairness of it all.

After a few minutes of rumination, I sat down at my desk and got to work. I drafted a memo to the rest of the C-suite and our division presidents to let them know I would be focusing on AI and could use their help. To the extent that I could, I blocked my calendar to be able to prioritize this, and, in a separate email, I let my executive assistant know about the importance of keeping that time free.

Eventually, I packed up my laptop to leave. The elevator ride to the parking garage gave me twelve floors of relative quiet to replay Richard's words over and over.

You have until next quarter to show us a real AI strategy. Otherwise, we'll need to make a leadership change.

The drive home took longer than usual, but I welcomed the delay. I needed time to figure out how to tell David what had happened without worrying him. He'd been so supportive when I took this job two years ago, even though it meant uprooting our family from Boston. The kids had adjusted well—better than I had, to be honest—but David had put his policy consulting business on hold, taking an adjunct position lecturing in ethics and public policy at Emory that he enjoyed but wasn't the same as running his own firm.

I pulled into our driveway just after 6:30 PM. Our home glowed warmly against the darkening sky. Through the kitchen window, I could see Emily at the island counter, probably working on her science project. Jake would be upstairs, pretending to do his homework while not-so-secretly gaming. And Lily would be on FaceTime with her Boston friends.

David met me at the door, dish towel over his shoulder. One look at my face and he knew.

"That bad, huh?" he asked, pulling me into a hug.

"Worse," I said into his shoulder.

"Dinner's almost ready," he said, steering me toward the kitchen. It was my favorite part of the house—open and warm with copper pots hanging above the island.

Emily looked up from her notebook and squinted at me. "Mom, you look terrible," she announced with the brutal honesty of a sixteen-year-old.

"Thanks, sweetie," I said, dropping my bag on a chair. "Great to see you too."

"Did the board not like your presentation?" she asked, only half paying attention to me.

I caught David's eye over her head. He raised an eyebrow, a silent question. *How much do we tell her?*

"They want an AI strategy," I said carefully. "And they want it fast."

Emily nodded. "My history teacher thinks AI is going to be the most disruptive technology ever. We did a whole unit on it."

"Is that right?" David handed me a glass of wine. "Maybe I should have consulted you before the meeting."

"Probably," she said, with a wry smile. "I got an A."

The casual way she said it caught me off guard. As if by osmosis, she seemed to understand AI the way children learn languages. It was just another part of her world. I felt profoundly old. Or, at the very least, exhausted.

David and I moved to the living room, sitting down on the couch.

"So," David said, leaning back against the cushions, "are they really that serious about this AI thing?"

"Ninety days to create an AI strategy or they'll replace me." I took a deep breath. "Richard basically told me I'm obsolete."

"You know that's bullshit, right?" David's directness was one of the things I loved most about him. In my world of corporate diplomacy, his honesty was refreshing. "You're at the top of your game. You've delivered on every metric they care about."

"Except this one," I said. "And apparently that's all that matters now."

David ran a hand through his salt-and-pepper curls. At forty-seven, the gray made him look distinguished, which suited him.

"So what's your plan?" he asked. Not if I had a plan, but what it was. His confidence in me never wavered.

"I'm not sure yet," I admitted. It wasn't something I could say to most people. But with David, I could be honest. "Right now, I feel like I'm drowning."

"You're not drowning," he said, taking my hand. "You're treading water—you're figuring things out. At some point you'll have to swim to the other shore, but right now, you're just treading water."

"That sounds a lot like a euphemism for drowning," I said dryly.

Just then, my phone buzzed. It was a text from Paul: *Sarah, I'm sorry about today. Coffee tomorrow? I want to help.*

I showed David the message.

"Good," he said. "Paul's always had your back."

I nodded.

"You've got more resources than you realize." David took my wineglass and set it on the coffee table. "Now, should we call the kids down for a family council, or do you want to wait until you have more information?"

I considered it. Our family councils had been a tradition since Emily was little—a time to discuss big decisions and changes together.

"Let's wait," I decided. "I want to know more first."

I typed back to Paul: *8 AM?*

Setting down my phone, I leaned into David's shoulder and closed my eyes. For a moment, I could hear only our breathing and the faint hum of the refrigerator. The house settled around us. Upstairs, a door closed quietly. Someone ran water.

ACT 1

Chapter 1

Starting Over

I arrived at the coffee shop fifteen minutes early, needing space to gather my thoughts before Paul arrived. Six hours of restless sleep had left me feeling hollow and sluggish. I was moving a half-step behind myself, slower to react, and foggier than I could afford to be. My shoulders ached from tension. There was a dull throb behind my temples.

Paul arrived composed and unhurried, well-dressed as usual in a navy blue blazer. Our eyes met across the room. No smile, just a slight nod of recognition. He'd been my mentor for years. Now I wasn't sure if he was still my ally or simply the bearer of bad news.

"You look like you slept about as well as I expected," he said, his tone carefully neutral. One hand rested lightly on the back of the chair opposite me, as if unsure whether to stay or stand. "What do you want to drink?"

"A black coffee. Thank you."

"You got it."

Paul returned a few minutes later with our drinks. I wrapped my hands around the warm cup.

"So," he began, "rough night?"

I nodded.

"The first time I got that sort of ultimatum," he said, with a sort of fond recollection, "I don't think I slept well for a week."

"How comforting," I replied, taking a sip. The coffee was good, bitter, strong, and smooth. "Was that before or after they fired you?"

Paul's eyebrow rose slightly—the closest he ever came to showing surprise. "Point taken."

We sat in silence. The coffee shop was already busy, the morning rush filtering in and out with a sort of anonymous clockwork urgency that I found comforting.

"You didn't say much yesterday," I finally said.

"I said plenty before the meeting."

"As far as I know."

"Would it have helped?" he countered. He wasn't defensive, but he was firm. "Boardroom battles are won before they begin, Sarah. You know that."

I sighed. "You know, I keep running through every possible scenario. I've built a career on understanding things. Really, deeply understanding things. Marketing, supply chains, logistics—things that didn't make sense to me at first, but I learned them. Piece by piece. But AI?" I shook my head. "I don't even know where to start."

"And that scares you."

"Of course it scares me. Ninety days to master something that I don't know much about and is making headlines everywhere."

Paul leaned forward. "The board's urgency isn't coming from nowhere, Sarah," he said quietly. "I've been hearing rumors that Apex Shipping is developing something significant. Something they could use to pick off our key clients."

I felt a knot tighten in my stomach. "What kind of something?"

"I don't know. But you can't afford to spend three months planning." He paused, studying my face. "Sarah, your instinct is to master every detail before you act. It serves you well. But this time, perfectionism might kill us."

"So what do you suggest?"

"You may not know what to do, but you know how to learn," he said. "Start by talking to the people who are already doing the work. And do it before that voice in your head convinces you to spend weeks researching all the questions to ask them."

"You're not the first person to call me an overthinker," I muttered.

"Look, it's just a hypothesis," he continued, "but if AI is anything like the other major tech transitions we've both lived through—the internet, mobile—the leaders who fail won't be the ones lacking technical know-how. They'll be the ones who couldn't admit they needed help."

"All right," I said. "So then the question becomes: Who at FreightCo has the most AI experience? And who do I trust enough to seek out their opinion?"

"Right questions. Likely different answers."

Two of FreightCo's division leaders came to mind. "Mark Armstrong," I said, thinking it through. "The president of our Western Division. He's talking a lot about AI. And . . . Julie Guthrie. I haven't heard her specifically mention AI, but I've heard other people talking about her division. People want to work for her."

"Text them both," Paul suggested.

I hesitated, the resistance instinctive. What would I say to them? Tell them that if I didn't come up with an AI strategy in ninety days I'd be fired? No . . . but how should I bring them in on this? I wasn't sure, and I didn't want to act without being sure. It was the same steadfast, protect-myself-at-all-costs resistance that had kept me up most of the night.

"Sarah, c'mon, do it now." His tone was firm but warm.

"Okay, I'm doing it." I pulled out my phone and quickly texted each of them individually asking when they could chat.

Mark's response was instant: *Absolutely! Free this afternoon for a Zoom? 2 PM?*

"Mark's in," I said.

"Of course he is." Paul smiled faintly. He stood to leave. "Sarah. Don't worry. You'll figure this out. I believe in you."

I thanked him. After Paul left, I sat alone for a few minutes, collecting my thoughts. A half-formed, lingering idea turned over and over in my mind: Maybe the first step wasn't understanding AI. Maybe it was understanding why I was so afraid of not understanding it.

Just as I was about to leave, my phone buzzed again. The text was from Julie:

Happy to talk. We could do a video call, but you'll get more out of seeing things firsthand. Could you visit us in Austin? We'd love to show you around.

I stared at the message, surprised and intrigued by the quiet confidence it conveyed. Almost without thinking, I searched for flights. There was a 6 AM tomorrow to Austin. I hesitated—the old fear urging more preparation, more research. But this was a new era. If I was going to succeed, I had to start by rewriting my own personal rules.

I bought the ticket.

Chapter 2

Innovation Theater

Mark joined the Zoom call precisely at 2 PM, flashing the camera an eager smile as if he'd been waiting all day just for this moment. He wore a crisp white shirt, sleeves rolled up neatly past the elbows, showing a gleaming silver watch.

"Sarah! Great to see you." He leaned toward the camera, framed by an array of industry awards that hung on the wall behind him.

"Hi, Mark." I offered a polite smile, feeling preemptively tired from an interaction that had barely begun. "Thanks for making the time."

"Of course," he said smoothly. "So, what can I tell you?"

"I'd like to get clear on our current AI initiatives," I said, trying to sound open and genuinely curious. Paul's words echoed softly in the back of my mind. *Ask for help, Sarah.* "I saw your recent press release about our 'AI transformation,' and I wanted to learn more."

His smile widened, mistaking my neutral reference for praise. "Oh, you saw that? Great coverage, right? TechCrunch picked it up. We're way past the basics now—honestly, Sarah, we should consider rebranding to FreightCo.ai to really signal our transformation to the market."

"I'll need to think about that. Let's discuss it later. But for now, can you tell me more about the 'revolutionary AI implementation' mentioned in the article? What have we deployed?"

"Sure. We're diving deep into generative AI across the board. Customer service automation, report generation—the whole thing."

"And what are the results?"

"The pilots are promising."

"Pilots? As in, you're still testing them? That's not what I thought you said in the press release."

"Advanced testing," he said quickly. "We're very close to going live."

"What's holding us back?"

He shifted slightly. "Just minor technical challenges. These systems are incredibly sophisticated. Sometimes the AI generates . . . creative solutions."

"Creative how?"

"Well, occasionally the outputs require some refinement. But that's normal with cutting-edge technology."

He was avoiding saying something. "Mark, what specifically is preventing these systems from going live?"

"Most of the time, they're great. But . . . sometimes they make mistakes. We all do."

"What kind of mistakes?"

"Sometimes the customer service bot offers promotions that need adjustment," he said, visibly agitated.

I pressed him. "And the report generator?"

"It's learning. Sometimes it misinterprets things. Or hallucinates. But that's the nature of innovation, right? We're pioneering here."

"Mark, it sounds like these systems aren't ready. And you shouldn't have been so quick to tout them to the media."

His phone buzzed on his desk. He glanced at it, and I watched his expression change completely.

"Is everything okay?" I asked flatly. My patience was wearing thin.

"It's nothing," he said.

"Is it work related?"

"Yes. It's from Weyco Industries."

"What do they want?"

"They have some questions about their pilot," he finally admitted.

"What kind of questions?" I asked, glaring at him.

Mark hesitated, clearly hoping I'd move on. "They're asking for clarification on some features."

"Mark."

He sighed. "They're saying our AI recommended a promotional discount they don't actually offer. But it's just a training issue—"

"What exactly did they say, Mark?"

"They want to discuss the situation."

"That's not what your face looked like when you read that message."

Mark's eyes darted away from the camera. "They're frustrated, obviously. But we can work through this."

"Mark, what did Weyco Industries actually say?"

A long pause. "They mentioned reviewing their options."

"Reviewing their options meaning what?"

"Look, Sarah, every pilot has bumps. This is normal—"

"Mark. Are they threatening to terminate their contract with us?"

"They said they're considering it."

The silence stretched between us.

"Sarah, you have to understand, the media wants to hear about AI innovation. We need to have a vision and share it widely. We can't wait until everything's perfect—"

"To what? Lie about our progress?" The words came out harsher than I intended. I pressed my fingers against my temple, forcing myself to breathe.

Softening my voice, I continued: "Mark, press releases like this create expectations it doesn't sound like we can meet."

He slumped back in his chair. "I thought you'd appreciate the visibility."

This caught me off guard. "Well, yes, to an extent, I do. But more than visibility, I appreciate honesty. Our customers trust us. We can't break their trust. I'm not saying we shouldn't innovate. But maybe we need people checking our AI's work? Make sure it's not . . . being too creative?"

"That's not where the industry is headed," he muttered. "The industry is headed toward automation."

"I'm not particularly concerned with what the industry is doing." This wasn't entirely true, but it was what Mark needed to hear.

"What do you want me to tell the media, Sarah?"

"I want you to build something that works, Mark. The media can wait. Let's pause all AI-related press releases for the next ninety days while we figure this out." I closed my notebook. "I'm traveling tomorrow, but we should talk after I'm back in the office."

He nodded, looking chastened. "I understand."

"Thank you, Mark."

After we disconnected, I sat in the quiet of my office, staring at my notes. A few minutes later, I joined my next meeting, which felt refreshingly grounded by comparison. At least the finance team's numbers added up, even if they weren't revolutionary. But the familiar ache had returned to my shoulders, the same tension that had kept me awake last night, and with it came a creeping dread. Our most vocal AI advocate was actively damaging FreightCo's reputation while burning through resources and without anything to show for it.

The board had given me ninety days to figure things out. Mark had just shown me I couldn't afford to waste a single one.

Chapter 3

Big Pragmatic Problems

It was still dark when I boarded the plane to Austin. As I settled into my seat, my phone buzzed with a text from Paul: *Hearing rumors about Apex. Something big is coming in the next 24–48 hours. Learn fast.*

My stomach tightened. I forced myself to focus on what I knew about Julie Guthrie. We'd known each other for almost three years—ever since she'd moved up from Operations manager to run the South Central Division. She had switched from philosophy to operations management in grad school, and I got the sense she was adept in both by the thoughtful, measured way she listened and asked questions.

The flight was short but turbulent. Stepping out of the airport, I was surprised by the warmth of the morning air. It had been a while since I'd visited Texas. When I did site visits, it was usually to help fix problems, and things always seemed to be running smoothly here.

When I arrived at the building, Julie was waiting for me at the main entrance, holding two coffees. Dressed simply in dark jeans and a blouse, she smiled as I approached.

"I'm glad you're here," she said, handing me one of the coffee cups. "We're really excited to show you what we've been working on."

"Thanks for making time," I said. "I've been looking forward to seeing what you're doing."

"Of course," Julie replied, leading me through security. "We've got a lot going on here, but there's a method to the madness. We'll cover all of it, I promise, but I want to start by showing you a relatively simple solution we devised to a big problem. One that is not very sexy: contract reviews."

"Okay," I said, letting her guide me through the corridors until we reached a large room where about a dozen people were working at clustered desks. The space had an energetic, slightly disheveled quality—whiteboards covered with diagrams, monitors displaying colorful interfaces, empty coffee cups, sticky notes everywhere. There was the low, background, we-are-really-doing-this hum of collective enthusiasm. It reminded me a bit of being in college at Carnegie Mellon, both working on group projects and, perhaps even more so, of watching other people work late into the evening as I prepared snacks and coffee for them during my shift at the library café.

"This is Anita," Julie said, guiding me to a desk near the center of the space, where a woman in her thirties was working at a monitor. "She's part of our contracting team. Anita, can you tell Sarah what your workload was like six months ago?"

Anita looked up, pushing her glasses up her nose. "Oh, I was drowning," she said. "Routine NDAs, standard service agreements, basic amendments—they were eating up all my time. Meanwhile, the complex stuff that actually needed my expertise was piling up."

"Some of the work Anita was doing was necessary but repetitive," Julie explained. "So we asked ourselves: What if we could handle the routine stuff differently?"

Anita pulled up a dashboard on her screen to show me. "The AI takes a first pass on everything, flagging specific sections in each of the documents that need reviewing based on our risk parameters. I get to focus more on reviewing genuinely complex matters where my judgment is needed."

She opened a contract. The system had flagged several sections with color-coded highlights.

"This red highlight means the AI is pretty sure there's an issue," Anita explained, pointing to a warranty clause. "Yellow means there might be an issue. Green means it matches our standard language. I still review all of it, but the color-coding is helpful in directing me to the areas that require the most scrutiny."

Anita hovered over a red clause, typed in a command, and the AI model suggested replacement text.

"I also review the text it generates," Anita explained. "I'm ultimately responsible for the output, so I still review all the work as I would my own."

"That's important," Julie added, "because generative AI gets things wrong. Generative AI models string together words and sentences based on statistical patterns they learned when they were trained on massive amounts of text. When Anita prompts this model, it doesn't really 'understand' her. Not in the conventional sense. It is able to produce what is *likely* to be a sufficient or correct answer, but when we first tried this, it was disappointing. The answers felt pretty canned. But we kept iterating and tried connecting it to our contracts database through a process called retrieval-augmented generation, or 'RAG,' so it can inform its responses with existing language specific to our clients, and that really helped."

"Impressive," I said.

"I suppose," Julie replied. "But there's always room to be wrong. For instance, Anita's intent and the AI's interpretation of her prompt might diverge. Or clauses in the documents in our contracting database might conflict. See what I mean?"

"I do," I said. "But it still seems impressive. What would you say its impact is?"

Julie turned, gesturing to Anita with the quiet confidence that she'd be able to answer with aplomb.

Anita seemed to really think before answering. After a few moments, she began: "I'd say I get through about two times as many contracts a day now as I did before. It's particularly helpful because the system doesn't get tired, or, you know, distracted. It'll flag things for me I may not have thought to check as closely. And it can pull together contextual information for me so much more quickly than I could do by hand. But I also catch mistakes it makes. In particular, I know more about our clients than it does, and because of that context, I still catch plenty of things. For instance, last week the AI flagged a warranty provision as nonstandard. Technically it was right, but I knew that client always requests that specific language because of their regulatory requirements, so I overrode it."

"Basically, we're trying to design teams that get the most out of the technology while also leveraging the strengths and expertise of our employees," Julie said. "It's an example of what we call human-machine teaming."

"I really like that," I said, nodding. It made intuitive sense to me. "Yesterday, I spoke with Mark Armstrong, the president of our Western Division. Although his team has been pretty vocal about their AI initiatives, in reality, they're struggling to get off the ground with AI—in part because they're going for, in his words, 'full automation.' Have you shared this approach with him?"

"To an extent, yes," Julie said, choosing her words carefully.

"What does that mean?" I asked. This was the first time Julie had been anything but straightforward with me.

"I've tried to share our approach with Mark on two occasions," Julie admitted. "But both times, he seemed to be more interested in talking *at* me than he was in talking *with* me."

Across the room, someone coughed.

"I see."

Julie continued: "Let me be clear: Mark and I share an enthusiasm for AI. I'm not pessimistic about it. I use AI all the time. To take meeting notes. To help draft speeches. To create presentation outlines. Last week it helped me turn a rambling brainstorm into the outline for a strategy document in minutes instead of hours. But there are some major differences between our divisions and the approaches we're taking."

"Julie," I said, sensing the opportunity, "please tell me more about exactly that."

Chapter 4

Permission to Fail

"People tend to think AI implementation starts with technology," Julie began. "That's pretty normal. AI is a new and exciting technology, and it can do a lot of things. But starting with the technology is backward. We started by thinking about business challenges and asked ourselves if AI could help and came up with a range of potential use cases. The contract review system you just saw came out of that process. I'll tell you more about it. But let me start at the beginning."

Julie took a deep breath.

"When the first large language models came out, I felt totally overwhelmed. I literally lay down on the floor one day because my head was spinning thinking about where the world was headed with AI. But once that feeling began to pass, I tried to learn as much as I could about AI. I went to webinars and conferences. I talked to people in my network. I interviewed people on my team. We even brought in an AI professional services firm to learn from them."

"Were they helpful?" I asked.

"Yes," Julie said. "One of the first things we did was go through a series of exercises with them to assess our AI readiness."

"Like how prepared you were to do things with AI?"

"Exactly," Julie replied. "That exercise revealed one of our biggest gaps: data readiness. We had a lot of data, but having data isn't the same thing as having good data, or data that is organized in a way that can be useful or even data that is available on the right architecture. So we had to spend time cleaning up our data, which ultimately was a good investment because AI systems are only as good as the data they're based on."

"How long did that take?" I asked, thinking about our timeline.

"Months," Julie admitted.

I winced.

"But the contract review use case was worth it. Apart from saving us real money, it was also pretty low risk as long as we retained human review. Plus, it was a reversible decision: We could always revert to manual review. And, of course, Anita's loving it, which is great as I'd hate to lose her. But the most important thing we tried to do was create space to learn from success and failure. We realized early on that if people weren't comfortable admitting they didn't know something, or that they were wrong, or if they felt we might be creating ethical issues, they wouldn't be able to do their best work. They'd also be less likely to share those concerns, which would prevent us from trying new things, incorporating lessons, and improving."

My jaw tightened. Two days ago, the board had threatened to fire me for not having all the answers about AI. And here was Julie, celebrating when her team admitted they didn't know something. I both admired and envied her.

"Come on," Julie said, perhaps recognizing I was overthinking things. "Let me show you something else."

Julie led me through a series of corridors, past developers with screens full of code. "Our engineering teams are substantially more productive now working with AI coding assistants," she said, almost as an afterthought. "That's pretty straightforward, though. I mean, everyone's doing that."

I nodded, although, in truth, I hadn't been aware that *everyone* was doing that.

"But you should be careful—we found that these tools really help experienced developers. But for less experienced ones, they can actually amplify mistakes." She continued: "As you can see, we aren't opposed to adopting what works. But our strategy isn't just to 'keep up.' I mean, if you just focus on the shiny object like everyone else, how could you ever differentiate? You're just doing what everyone else is doing, with very little critical thought."

Soon we reached a control center. It was a dimly lit space that smelled of pine air freshener and had a view of the warehouse and multiple displays showing regional maps.

"This is where the rubber meets the road," Julie said, introducing me to Raj, explaining that he managed their custom AI-based route planning system.

"We use our route planner to manage deliveries for eight hundred trucks across five states," Raj explained, gesturing at a map dense with colored paths. "I can go in here and see near-real-time information about routes, whether they're ahead of or behind schedule, and what's in the queue. I can also test different scenarios to see how they might affect pickups, deliveries, and return trips. We try to keep delays to a minimum and reduce backhauling as much as we can."

Julie cut in: "We could try to optimize for cost savings, but we don't. Can you guess why?"

"Because, ultimately, it isn't what's best for business," I reasoned. "Optimizing for cost savings might increase delays and backhauling. Delays jeopardize customer loyalty and long-term revenue. And backhauling is a huge drain on resources."

"Exactly," Julie said, looking to Raj to elaborate.

"Right, so we found that in training our neural network—our AI system—with those reward signals, what we essentially trained it for is route *resilience*, or the routes that satisfy our constraints under the most probable conditions. You can think about it another way: 'Perfect' is brittle," he explained. "Construction can create delays or make some routes impossible. Storms too. Or, you know, customers might request different delivery windows. A lot of things can happen."

"Show her last Thursday," Julie said to Raj.

He entered the date. The map showed a massive storm system moving through Texas. "The system started rerouting trucks hours before the storm hit, projecting better alternative routes based on the forecasted path of the storm and historical weather delays."

"That's . . ." I paused, trying to find the right word.

"It's not magic," Julie said, reading my expression. "It's just pattern recognition and predictive analytics at scale."

Raj pulled up a different visualization that showed route patterns over time. "The system gets feedback from every delivery and learns a little bit more each time, so it's better the next time."

"But it didn't start this way," Julie continued. "These capabilities are only possible because we did the initial training on data we've been collecting for years. Some of it was just the by-product of how we ran things here. But two years ago, we began to view data as an asset, not just as a by-product. We started cleaning and cataloging it. We found ways to collect it where we weren't before. *That's* the real advantage," Julie said, getting more animated.

"Competitors could try to implement a similar solution tomorrow," she continued. "But unless they have the data to train the model, they can't do it. There's an old saying—the best time to plant a tree was forty years ago, and the second best time is today. Well, it's the same thing with data: We didn't start thinking strategically about data twenty years ago, but we started as soon as we realized we needed to."

I thought about Mark: For all of his rhetoric, he hadn't mentioned data once.

"Could we chat in private?" I asked.

Julie agreed, and we walked to a nearby conference room.

"I'm impressed by what you've shown me," I said, shutting the door behind myself. "I really am. I'd love to take what I've learned today and share it with a wider audience."

Julie didn't react to this in the way I thought she would. Instead of looking happy, she seemed to deflate a little bit, drawing in on herself.

"Talking about our work hasn't exactly been rewarding," she said.

"What do you mean?"

"Well, for one thing, someone at Apex must have caught wind of what we're doing, because they've been trying to poach my team. Offering salary increases—not just to our engineers, but also to people like Raj who understand how to operate these systems. They must know what we've built and want to shortcut their own development."

"What?" I said, surprised and more than a little embarrassed that Apex seemed to know more about what one of my divisions was doing with AI than I did. "Are people leaving?"

"Just a few," she admitted. "I think everyone else will stay because they value what we're building and because they appreciate our team culture. And, from all I've heard, Apex isn't a great place to work," she added. "Also, I really feel that other teams in our division have felt threatened by what we're doing here. I think they worry that AI might take their jobs, which isn't remotely a part of my philosophy. In any case, I haven't been broadcasting much of what we're doing because I don't want to tip off our competitors and don't want to worry our team."

"Julie," I said, and touched her lightly on the shoulder. "I'm sorry I haven't come to talk to you about AI until now. I didn't know what I was missing. And that's my mistake." I continued: "But I'm going to need your help. The board wants a comprehensive company-wide AI strategy in ninety days. Rumor has it that Apex is going to make a big AI announcement soon. And Shop & Buy is reconsidering their contract with us. There's a lot at stake here. And if I'm being honest with you, I felt totally overwhelmed coming here today. But seeing what your team is doing—seeing what you've built—it gives me confidence that we can do this. Really," I said. "I'm impressed."

Chapter 5

Not Just Behind

After I said goodbye to Julie and her team, I made my way back to the airport. It was getting late in the day, and Apex hadn't made their announcement yet. At this point, I wasn't all that worried. Julie's team had some impressive capabilities, and I didn't have any reason to believe Apex would be able to leapfrog what I'd seen.

I settled back into my seat as the plane roared down the runway and then lifted off into the relative quiet of the sky. Below, Austin's contours shrank into the distance, giving way to patchy countryside punctuated by suburbs and then, eventually, small towns. I opened my laptop, trying to organize what I'd learned.

I created a new document titled "AI Strategy: Initial Thoughts" and began typing:

Core Principles:

- *Start with real business problems, not tech possibilities.*
- *Build data foundations today for advantages tomorrow.*

- *Design human–machine teams instead of replacing people.*
- *Learn continuously by making experimentation safe rather than expecting perfection.*

Just then, the flight attendant came over and offered me a beverage. It had been a long day, so I saved the document and closed my laptop. The path forward was becoming clearer, even if the magnitude of change was daunting. I didn't have Wi-Fi on board, and I was spent from thinking. The best thing I could do right now was get a bit of rest.

I woke up as we began our descent toward Atlanta. A few minutes later, we were on the ground, and my phone was buzzing nonstop.

Paul's text was the first one I read: *Are you seeing this?*

The next one: *This is bad news.*

I clicked through a few more messages until I found the headline:

Apex Shipping Partners with Microsoft to Launch "First Fully AI-Native Logistics Platform"

I hastily skimmed the article. Automated routing that promised 99.9 percent optimization. Predictive maintenance that would eliminate breakdowns. Customer service powered by AI that could handle complex challenges. Apex wasn't just implementing AI tools—they claimed they were rebuilding their entire operation from the ground up.

The moment I was off the plane, I called Paul.

"Sarah," he answered immediately. "I assume you've seen the Apex announcement?"

"I was on a plane and just read it," I said, walking faster now, my heels clicking on the terminal floor.

"It's worse than you think," he said. "I just got off the phone with Jim Hartley at S&B. They saw the Apex announcement, and they're already asking questions about our AI road map. They want reassurance. Soon."

"How soon?"

"Soon enough that Richard's holding an emergency board call tomorrow morning."

I stopped walking. Travelers passed me in the terminal, but I barely noticed them.

"Sarah? Are you still there?"

I swallowed. "M-hmm."

"There's something else. My phone has been blowing up. Everyone's spooked. They're all asking the same question: Can we compete with Apex?"

"What are you telling them?"

"That we're working on it. But, Sarah, platitudes won't hold them for long. And there's more—I'm hearing rumors about talent flight. Engineers getting offers. The kind of offers that are hard to refuse."

Shit. "Yes, I heard that too," I admitted, unsure if I should tell him I'd only *just* heard that from Julie's team.

"You did?"

"I did."

"Well, we're not just behind—we're at risk of being hollowed out." Paul sighed. I imagined him rubbing his temples on the other end of the line. "Please tell me you've learned something from Austin."

"I have," I said. "I learned we're already doing significant things with AI. Julie's team is way out ahead. My guess is whatever Apex says they're doing isn't actually as impressive as what her team is doing."

"Really?" Paul said, sounding cautiously optimistic. "Why haven't we heard more about what her team is doing?"

"From now on you will, believe me. She's been working methodically to produce results. Maybe she could've engaged in more self-promotion, but that doesn't seem like her style, and I can't say she's gotten positive feedback for sharing what she's doing."

"What do you mean?"

"We're a competitive company, Paul. People get threatened by results."

I let the silence hang between us. I didn't excuse it. I'd faced my own share of obstacles as a high-achieving woman at FreightCo, and Paul knew that.

"I'm going to learn from Julie and her team," I finally said. "This is not easy. She doesn't have it all figured out, but I—we—can learn from working with her. I think we can put something together that will win."

"Well," Paul said matter-of-factly, "you have my support. And I imagine you'll have access to more resources too. It isn't just your job on the line anymore. It's all of ours."

ACT 2

Chapter 6

A Moment of Crisis

Richard's assistant had scheduled the emergency board meeting for 7 AM. I logged into the videoconference two minutes early, sitting stiffly as I watched rectangles populate the screen. Richard appeared last. Behind him, I could see the New York skyline, still dark in the early morning.

"Let's get straight to it," he began. "Apex Shipping isn't just talking about AI anymore. They've partnered with a major tech company, and they're signing clients. Our stock closed down fifteen percent yesterday."

"And Shop & Buy?" Paul asked, adjusting his glasses.

"They called. They want to understand our AI capabilities. Soon." Richard's jaw tightened. "Our biggest client isn't sure we can keep up."

Multiple people moved to speak, but Diane unmuted first: "There are a dozen AI start-ups in logistics we could buy today. We should make an acquisition and get capabilities immediately."

"Acquire what capabilities, exactly?" Paul countered. "We

don't even know what we need. Panic buying at inflated valuations isn't a strategy. At least, it isn't a good one."

"It's better than doing nothing," another board member jumped in. "At least we'd have something to announce."

Michael, usually one of my supporters, spoke up: "Maybe we should stay entirely internal. Double down on our own development."

"With what expertise?" Richard shot back. "Sarah, we gave you a mandate."

"Three days ago," I said.

"I realize that. But you should've been preparing for this for years."

I sat with that for a few seconds before responding.

"I've spent the past few days learning from our own divisions," I said. "I was in Austin yesterday. Julie Guthrie's team has an impressive AI solution that's already saving us money."

Diane interjected: "Surely that isn't enough. Apex is promising transformation, and what, we're talking about cost savings?"

The board debated among itself for a few more minutes, gradually coalescing into three factions: acquire something, anything; stay internal and hope; or bring in external experts to save us.

Paul was the only one who seemed to rise above the fray: "Sarah's right to try to learn from our own people first," he said carefully. "But we might also benefit from an external perspective."

"We can't afford to waste time learning," Richard said. "We need results."

"Those aren't mutually exclusive," I said. "Learning is a result. Full stop. And we need to learn if we want to create a compelling strategy."

Diane began to respond, but Richard cut her off: "Sarah, we need a plan we can share with clients to give them the confidence to stay with us. We can't afford ninety days anymore. You have thirty."

"I understand."

"You understand?" he asked. My straightforward answer seemed to have caught him off guard.

"Yes."

"Okay, then. Good luck."

Board members began dropping off the call. Paul stayed on.

"Don't leave yet," he said once we were alone.

Yes? I thought. *What is it you want to tell me, Paul?*

He leaned back in his chair, suddenly looking tired. "Can I be honest with you, Sarah?"

"Please."

He glanced off-screen for a moment, as if checking that his door was closed. When he spoke again, his voice was lower. "The board doesn't understand AI either. Not really. And that includes Richard."

My face must have given away how I felt, because Paul acknowledged my surprise.

"Yes. I know it's hard to believe. But we're all reading the same articles. We're seeing the same headlines. We're feeling the same pressure. We don't know." And then he added, almost apologetically: "But we aren't saying it."

Anger rose in my chest. "You're telling me the people who just threatened to fire me don't understand what they're asking for?"

"I'm telling you there's a difference between recognizing a threat and knowing how to respond to it."

"Then what *exactly* am I supposed to deliver in thirty days?"

"A plan that makes us feel confident we won't be left behind." Paul's eyes crinkled at the corners—it wasn't quite a smile, but it seemed to convey a bemusement that I could only imagine he had gained from living through these types of crises before.

"Look, people fear what they don't understand. And we don't understand AI, not really. That goes for me, Richard, and the rest of the board." He paused, choosing his words carefully. "I mean, we know AI is going to reshape every industry. And we're asking questions. I think they're the right ones. But we don't have the answers any more than you do. And we shouldn't. That's not our job."

I stared at his image on my screen.

"The boards who pretend to have all the answers are the ones that scare me. Our job—our real job—is to do two things: spot the danger and hold you accountable. Your job is to figure out what to do about it."

"Why didn't you tell me this before the call?"

"What would you have done with that information?" His gaze was steady through the camera. "Would you have spent the night crafting the perfect rebuttal to Richard? It would've been a distraction."

"You still could have defended me."

"I did defend you." An edge entered his voice. "I defended you on the call today. And I've been defending you this whole time. That's why you had ninety days to begin with. And it's why you have thirty days now instead of zero."

I hesitated. "Who wanted me gone?"

"I'm not going to tell you. But I spent most of last night shutting down the movement that was building to fire you this morning. A CEO change to appease investors and clients—that was their solution. Thirty days was the best I could negotiate."

My stomach tightened. "I had no idea." I should've said something else—maybe "thank you"—but my head was spinning.

"You can do this, Sarah," Paul said. "I wouldn't have fought for those thirty days if I didn't believe that."

He logged off, leaving me alone in the quiet of my home office. The morning sun was fully up now, streaming through the windows. Somewhere in the house, I could hear David making breakfast. I closed my laptop and went to find him. If I was going to figure this out, I couldn't do it alone.

Chapter 7

Family Council

David was in the kitchen making scrambled eggs. He was looking down at the pan and didn't notice me until I spoke up.

"We need to talk," I said.

"What's going on?"

"I have thirty days instead of ninety."

"Shit. That's insane."

Yes, it is.

Before I could say anything else, Emily walked into the kitchen. Jake and Lily followed closely behind.

"What's up, Mom?" Emily asked.

"Let's get our breakfast and then I'll explain."

When we were all settled around the table, I began. "So, as you know, I've been under some pressure at work lately. Earlier this week, I found out I have ninety days to do something pretty difficult."

"What's that?" Jake asked.

"Well, I have to convince our board I have a plan for how we'll use AI effectively. I'm making some progress, but this morning, the board accelerated my timeline: Now I only have thirty days."

Emily looked up from the table. "Mom, are you going to lose your job?"

The directness caught me off guard. David shot me a concerned glance.

"It's possible," I admitted. "But I'm going to do everything I can to prevent that."

"What would happen if you did?" Emily asked, her voice carefully neutral.

"We'd figure it out," David said. "We've made bigger moves before."

"I don't want to move again," Lily said. "I just made friends here."

"We're not moving," I said quickly. "Whatever happens, we're staying in Atlanta. But I may not be around very much over the next thirty days. I need to go back to Austin. And I'm not sure how long I'll be there yet."

Jake shifted in his chair. "Can we help?"

The question surprised me. "What do you mean?"

"I don't know. Research? We use AI for homework all the time."

"That's cheating," Lily said.

"It's not cheating. Some of my teachers have us use it," Jake countered. "To learn."

"*Some* of your teachers," Emily said, with the knowing condescension of a big sister. "Not all of them."

"It's not my fault that I'm good at using it! And besides, am I supposed to not use it even when everyone else is?"

Emily looked at her father. "Care to weigh in, Dad?" David probably had these kinds of conversations every day with his students.

"No, not right now," he said. "We need to get going to school."

After the kids went upstairs to get ready, David and I returned to the kitchen, and I helped with cleaning up.

"How long do you think you'll need to be in Austin?" he asked.

"At least for the rest of the week. I might need to stay through the weekend. Can you manage with the kids?"

"I can," he said, although I could hear some hesitancy in his voice. "But, Sarah, do you think they'll really fire you?"

"I'm honestly not sure," I admitted.

"Well, do you have a plan?"

"I'm working on it. I know I need to learn as much as I can as quickly as I can. Julie mentioned in her approach that she sought advice. I've already reached out to some of my friends running companies. And I'll ask Julie if she can bring in the consultants she worked with so I can talk to them too. Heck, I think I'll even invite Mark."

David raised an eyebrow. "Mark? I thought that guy didn't know what he was doing."

"That may be so. But he leads a division, and he's trying. I think his perspective could be important. At the very least, if we bring him along now, he's more likely to be bought in and could help us implement more quickly once we have a strategy. I think if we leave him out, he'll be harder to convince later."

"Savvy," David admitted. "But how are you going to make sure Julie doesn't feel threatened?"

"You mean by Mark?"

"Well, by this whole thing. You said last night she hasn't had the best experience engaging with FreightCo people outside of her division around AI. I get that you're coming in to learn from her, but I could see how, from her perspective, she might not know how to interpret your intentions, you know?"

"That's a good point. But I think that if I'm totally honest with her and I admit that I don't know what I'm doing and that I really need her help, she'll be fine."

"Of course she will," David said, drying the last plate.

I pulled out my phone and started typing. First, a message to Julie:

I need to come back to Austin. I can meet tomorrow morning. I know you're busy, but this is important—and a mandate from the board—and I really feel that I can learn from the work you're doing. Could you clear your calendar for tomorrow and Friday?

Her response came quickly: *Of course. We're all ready to help.*

I hesitated, then added: *Thank you. I think I am going to invite Mark Armstrong. I think he can learn from you, too, and if I'm there, he'll be more inclined to listen.*

The three dots appeared and disappeared twice before her reply: *Interesting choice. But yes, I trust your judgment.*

One more thing, I typed. *That AI consulting firm you mentioned—could they join us?*

Yes. I'll reach out to them. They have been great.

"She's okay with it," I told David.

"You're going to figure this out."

"You sound pretty confident."

"I've watched you climb steep learning curves before." David squeezed my hand. "Okay, I'm going to get ready for work."

"I can take the kids," I said.

Alone in the kitchen, I texted Mark: *Change of plans. I'm going to spend the next few days in Austin with Julie Guthrie's team. I want you to join us.*

His response took several minutes and came as I was getting the kids into the car: *I have meetings here.*

Please reschedule them if at all possible. This is the board's top priority. I need your input.

When?

I'm going to meet with Julie tomorrow morning in Austin. Get there when you can.

I pocketed my phone and drove the kids to school. Then I headed into the office. Later in the day, after a few meetings, I scrolled through my contacts and picked out a few friends who might be good to talk to: Dante, who ran a SaaS company and was pretty advanced when it came to new technology. Noah, who led a pharmacy chain and seemed open to new technology, at least conceptually. And Maya from a financial services firm, who always seemed to be focused on risk.

Quick question, I typed to each of them. *How are your boards reacting to AI? Mine wants results in 30 days. Would love to compare notes.*

Dante responded first: *30 days? Mine wanted an AI strategy yesterday. Happy to talk.*

Then Noah: *Board thinks it's hype. Maybe it is. But don't have $$ to even find out. Things are rough.*

Maya: *Risk and compliance nightmare. Happy to share what we're doing, but it isn't much. I'm on vacation with my family this coming week. Can give you a call when I'm back.*

I responded to set up conversations with each of them. It felt good to know that I wasn't alone.

Later on, after a family dinner as I was settling into packing for my return trip to Austin, my phone buzzed one more time. It was from Paul: *Heard you're bringing people together. That's good. Real change never happens alone.*

Chapter 8

AI Is Different

I ducked into the restroom before entering the conference room. Briefly, I considered splashing cold water on my face to wake up, but that would make my makeup run. I smiled at myself, hoping to inspire some positivity through sheer force of will. But I was tired, and it showed. Circles had formed under my eyes, which stung from the dry plane air. I rummaged through my bag to find my eyedrops and used them. After rubbing my eyes, I blinked a few times. I rolled my shoulders to try to loosen them up and reduce the aching. Then I stood up straighter, adjusted my blouse, took some slow, deep breaths, and looked myself in the eyes. *Come on*, I thought. *You can do this*. I tucked a loose strand of hair behind my ear. *You can*—

I jolted as another woman walked in. We briefly exchanged pleasantries as I grabbed my bag and suitcase and left.

I stood in the hallway for a few seconds to collect myself. Then I took another deep breath and walked into the conference room. Julie was engrossed in conversation with two people I didn't know, but I figured they were the consultants. They turned to me as I entered the room.

"Allow me to introduce Sarah Foster, our CEO," Julie said. "Sarah, this is Oliver and Ava. They're with Guidebridge.ai."

"It's great to meet you, Sarah, and really appreciate the partnership with your firm," Oliver said, introducing himself as the cofounder and CEO. His handshake was firm but not domineering. Dressed casually in a rumpled linen button-down and carrying himself with an ease that suggested he wasn't easily bothered, Oliver reminded me of Paul.

"We've enjoyed working with Julie and her team," Ava said, shaking my hand and explaining that she led their AI Strategy team. "She's filled us in on the situation. We're excited to help."

Ava was noticeably younger than Oliver but followed up on his introduction as an equal in the business. Like him, she was amiable and direct.

"Thank you for being here, and on such short notice," I said.

Ava waved away my comment cordially. "Oh, it's not a problem at all. We're based in Austin but most of our clients aren't, so it was easy."

Not just a local firm, I noticed. This seemed like a good sign to me.

"How did we become one of your clients?" I asked.

Julie exchanged glances with them, a slight smile appearing on her lips.

"I'll tell you more about that later, Sarah. But the short of it is that I had a—let's say *less than ideal* experience with a large firm with a global brand, and after that, I conducted a comprehensive search for a new partner. Guidebridge.ai scored the highest on our selection criteria."

I nodded. And then, suppressing the feeling that I needed to be the one to drive the conversation, I asked, "Where do you think we should begin?"

"Most executives want to jump straight to use cases—"

"Believe me," I said wryly, thinking of Mark, "I know."

Oliver chuckled. "Right. But that's exactly why so many AI initiatives fail. AI isn't like other technologies. We thought it would be best to begin by baselining on the ways AI is different from traditional software. I'm sure you understand traditional software—how it works, how to manage it. A lot of people implicitly anchor on traditional software as their mental model of AI. It's a mistake because the two are different in both big and small ways—and those differences can help you understand when and where traditional approaches don't work. But by starting with what you know and showing where AI diverges, you can build a mental model that will quickly help you navigate AI. We realize that it might feel like a lot to take in at first, but we think it's the best way to get you up to speed quickly given the time-sensitivity of the challenge we're facing."

"Okay," I said, pulling out my notebook. "Let's do it."

"All right, then." Ava pulled up a simple table with three columns that contrasted various aspects of Traditional Software with AI/ML.

"So, traditional software is logic-based," Ava began.

"I'm sorry," I interjected, pointing to the top of the right-most column. "ML is machine learning?"

How AI Is Different from Traditional Software

Aspect	Traditional Software	AI/ML
User Experience	Deterministic	Probabilistic
Adaptability	Requires discrete updates	Able to be updated using new data
Performance Basis	Performance based on code	Performance based on code and usage context at time of inference
Decision-Making Basis	Logic-based (more transparent and interpretable)	Data-driven (can be a black box)
Development Approach	Linear	Iterative
Project Management Approach	Waterfall; traditional scrum; etc.	Wayfinding (hybrid agile)
ROI	Focused on efficiency gains, sometimes new revenue	Focused on efficiency gains, new revenue, and a valuation premium for tech leadership
People	Hire experts; talent impact is typically linear	Hire learners; talent impact is exponential
Culture	Plan and execute culture; psychological safety is desirable	Learn and adapt culture; psychological safety is needed to encourage experimentation

She nodded.

"And what's the difference between AI and machine learning?"

"That's a great question," Oliver said. "The short answer is that AI is an umbrella term. Machine learning is a subset of AI, meaning all machine learning is a type of AI, but not all AI is machine learning."

"That's a helpful distinction."

Picking up the thread, Oliver added: "Most of the AI we're all reading about today is machine learning, but not all of it. For now let's just simplify it and call it AI. And we like to say AI isn't a tool—it's a toolbox. We can unpack that toolbox and the terminology with you, but I'd suggest we cover these key differences first."

"Let's start with the 'user experience' row," Ava continued. "Traditional software creates a deterministic user experience. If this, then that—2 + 2 = 4. It's heavily rules-based and interpretable. But AI and machine learning models create a probabilistic experience. They're essentially giant statistical pattern matchers. They find the 'best solution' for particular inputs based on data they were trained on, and for the user, that often feels like casting dice and seeing what you get."

She leaned forward. "Even language models that produce fluent text—they're trained on massive amounts of language data, and because of that, they can predict probable words across many, usually impressive, sentences. There's certainly utility in that. But because performance depends not just on the code but on the usage context at the time—meaning the prompt you use and, in some cases, the data the model is referencing, like a document—you might not get the same answer twice."

"Plus, traditional software requires discrete updates to change," Oliver added, pointing to the row labeled "Adaptability." "But AI models can be updated using new data—they're more adaptable."

"In some contexts," Oliver added, "this is really useful. Particularly contexts in which you have high-quality training data, have accepted some amount of variation in output, and have included human review—for instance, in proposing a variety of route options based on historical data. But in contexts where the margin for error is low and you've bypassed responsible oversight, it can be bad."

"Like a customer service bot?" I asked, thinking of Mark's fiasco.

"Exactly. Typically, with customer service, there's some underlying logic, like a decision tree. If a customer says X, then Y. But language models aren't following those rigid rules, so while they'll provide an answer, you can't guarantee it'll be right every single time. And when it's wrong—maybe it's only one percent of the time, but if there's no human to intervene, you might end up offering a customer a deal that doesn't exist, for instance, or respond to a comment unrelated to the business with something inappropriate."

"That's fascinating—and concerning. But what am I supposed to do about that? Aren't there guardrails you can put on such a system?"

"Well, for one thing, you can't expect the same development timelines or project management approaches you're used to."

"Such as?"

"AI development is overwhelmingly iterative. Progress is nonlinear. That means you need to manage AI development differently too. The traditional waterfall approach to project management doesn't work, and even agile needs to be adapted for the unique challenges of AI."

"Adapted how?" I asked. Agile had always felt cutting-edge to me.

"We use an optimized agile framework we call Wayfinding—it's specifically tuned for AI and ML projects. While traditional scrum might push you to start building on day one, AI projects need more time up front for hypothesis testing and data exploration. But we'll dive into that later. There are a few more fundamentals to understand."

I nodded, and wrote down *Wayfinding*.

"AI models tend to be more adaptable over time," Ava continued. "You can improve them by giving them access to new datasets, by providing human feedback on outputs, by fine-tuning them for specific tasks, or by connecting them to your company's knowledge bases—all without changing the underlying 'rules.' The rules stay the same regardless of the type of AI: The model is trying to find the statistical sweet spot for a particular input or prompt."

"Like finding an average?" I asked, trying to follow.

"Essentially, yes," Oliver said. "Now, in terms of measuring ROI, investments aren't just about driving growth or EBITDA. AI investments can fundamentally change a company's business model or upend competition in an industry, so they can also impact how a company is valued overall—both positively and negatively."

That'll be an important point to bring to the board, I thought.

"Lastly, in terms of people and culture—and you may have already been picking up on this with the time you've spent with Julie's team—to be successful with AI, you need to be intentional about building a culture where people feel safe admitting when they don't know something, where they can take risks, and where they share learnings with each other. When you're hiring, you want to hire people who will tend to behave like that."

"We actually adopted a framework from an author named Patrick Lencioni for this," Julie added. "His 'Ideal Team Player' model. We look for people who are hungry, humble, and people-smart."

"That framework has been invaluable," Oliver agreed. "Especially the humble part. AI work requires admitting what you don't know."

"M-hmm." The irony stung. Just three days ago, the board had threatened to fire me for not having all the answers.

"You should also keep in mind that quality beats quantity in hiring AI talent," Ava continued. "There aren't many experts, and the real experts don't have a linear impact. Ten times the team size doesn't equal ten times the speed. In AI, one top performer can be incredibly impactful—in particular, because of how they can contribute to the Wayfinding process. In this context, the traditional consulting model of putting a bunch of junior people on a project under the leadership of a senior manager—very likely someone who doesn't really understand AI—completely breaks down."

"That's fascinating." I shifted in my seat, feeling the weight of everything they were sharing. "Honestly, I feel like we could spend all day just talking about any one of these concepts."

Julie studied me for a moment. "Tell you what, Sarah. Let's keep going for now. But how about I arrange for you to talk with people from my leadership team about how we approached each area? Our Head of People about hiring strategy, our CTO about architecture, that sort of thing."

"That's a great idea. I'd really appreciate that."

Just then, my phone rang. My friend Noah, the CEO of Greentown Pharmacy. I checked my watch and realized I told him I'd call him at noon.

"Sorry, I need to take this," I explained.

"Not a problem!" Oliver said. "You take that. We've arranged to bring in lunch."

"Perfect. And thank you," I said, stepping into the hallway.

I closed my eyes for a moment, then answered.

Chapter 9

The Reality Check

Noah," I said, "it's been a little while. How are you?"

He sighed. "I've been better, Sarah."

I looked over my shoulder at the conversation continuing in the conference room, and, keeping my voice down, asked: "What's going on?"

"You wanted to talk about AI, right? Well, that makes two of us. Unfortunately, I can't say the same thing for my board."

"What do you mean?"

"There's no consensus, but overall they think it's snake oil, Sarah. That it's not ready for prime time. That I should basically ignore it for now."

"*Really?*" Had he pitched his board on AI the same way Mark had pitched me? "Why?"

"We tried bringing in consultants six months ago. Some boutique firm that claimed to be fricking AI experts." His voice had a bitter edge. "Turns out they'd just slapped ChatGPT onto their existing services and raised their rates. We burned through 100K for a 'transformation road map' that was basically a list of AI tools anyone could google."

"Ouch." *So that was it.*

"Yeah. So now when I tell the board it's important and we need to invest in understanding how to adapt, they just hear 'I need more money' and they imagine empty promises."

"I'm sorry, Noah. But not all consultants are like that—"

"I know," he cut in. "But try telling my board that. They won't approve another dime for AI exploration. We're experimenting with free and inexpensive tools, but I'm not sure what I'm going to do. I'm really worried about this. I mean, we all saw what the internet did to video and book stores."

"If it makes you feel any better, my board threatened to fire me this week." I chuckled in a gallows-humor sort of way.

"No shit?" I could hear the surprise in his voice. "Why's that?"

"Because I don't have an AI strategy. And, well, I'm sure you've seen the news about Apex."

"Figured that's why you texted me."

"Yeah."

"Figured you might want some advice."

"Yeah, in a sense. I wanted to know what you were doing with regard to AI, but it seems our boards are 180 degrees apart."

"Well, I'm sorry I can't be of help." Then he added: "What are you doing?"

"Right now, I'm in Austin learning from a division president of mine. She's great. And she's working with these consultants who seem to really know their stuff. My board has given me a month to figure things out."

"Sarah?"

I turned around, and there was Mark, a briefcase slung over his shoulder and a suitcase in hand.

"Hold on a second, Noah." I lowered the phone. "Mark. You made it."

"Cleared my calendar."

"I'm glad you're here." Then, gesturing to the conference room, I said: "Everyone's inside having lunch. I'll be in shortly."

Mark looked past me through the glass wall.

"Who are they?" he asked, referring to Oliver and Ava.

"Why don't you introduce yourself? I have to get back to this call."

Mark nodded and walked past me into the conference room.

"Sorry, Noah. Another member of my team just arrived."

"No worries. I was just thinking about your situation. And, you know, a month isn't so bad. I mean, you have money to invest, right?"

"I assume so. It's our biggest priority now."

"You know, you should consider yourself lucky, Sarah. It could be worse."

"I see your point. I hadn't thought about having to fight my board to invest in AI." And then I added: "I'm sorry to hear about what you're up against." I offered to share anything I learned with him in case it helped him build a case with his own board.

After I hung up, I went back into the conference room. Mark stood, arms crossed, talking to Oliver while Julie and Ava were sitting, eating lunch, and looking over nervously at the pair of men.

"I see everyone's met," I said, grabbing a sandwich.

Oliver nodded. "Yes, we have." And then, excusing himself from the interaction: "Looking forward to working with you."

"Likewise," Mark said curtly.

"Why don't we jump back in again?" I suggested.

"All right," Ava said, getting the conversation back on track and adjusting to sitting at the front of the room again. It took her a minute to get things set up. Eventually, she pulled up a circular diagram. Three overlapping circles—Strategy, Operations, Engineering—with arrows connecting them.

A Framework for AI Transformation

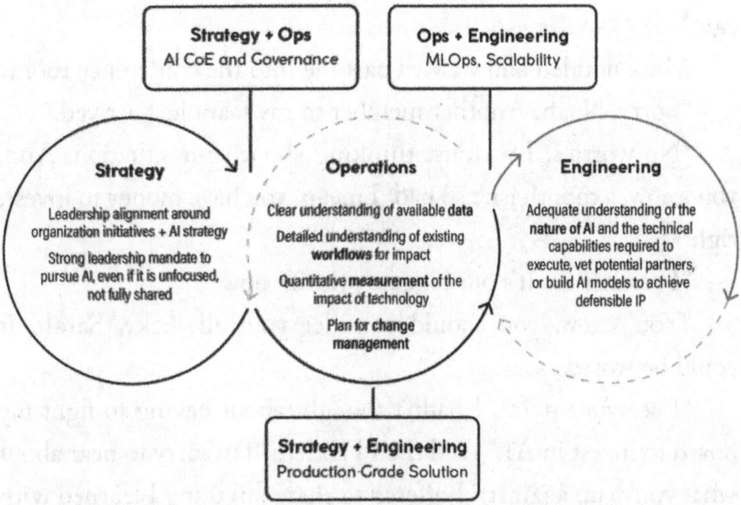

"We covered a lot this morning about how AI is different from traditional software. So the next thing we thought it would be worth showing you is a framework that contextualizes those things within major components of a company. Another way to think about this is: How should we do things differently as a result of AI?"

"This is our proprietary AI transformation methodology," Oliver explained, gesturing toward the diagram. "Think of it as a flywheel where each component reinforces the others."

I studied the diagram, thinking about Richard's ultimatum, specifically as I read "strong leadership mandate to pursue AI, even if it is unfocused, not fully shared." His words from Monday's board meeting echoed in my head: *You have thirty days to show us a real AI strategy.*

I thought, *Well, at least we've got a strong mandate.*

"Please," I said, "walk us through this."

"First, we find that with AI strategy, leadership alignment is key." Oliver stood, moving closer to the screen. "The initiatives your organization is taking on must directly reflect and reinforce the company's overall business strategy. So, for instance, if your strategy is to become the most responsive logistics provider, your AI initiatives should reinforce your focus on speed—routing that prioritizes meeting rapid deadlines, sufficiently fast customer updates, and so forth."

"Second, the mandate to pursue AI is foundational, even if the specifics aren't clear yet," Ava added. "Without executive—ideally CEO-level—sponsorship, every experiment that fails becomes fodder for skeptics. And developing AI that actually delivers value takes time. It's inherently messy. You're going to have false starts, dead ends, models that work in testing but fail in production. If you're trying to do it in an environment that doesn't tolerate failure, you'll fail."

"Well," Mark said, looking at me. "We certainly have board support for AI."

"That's ideal," Ava responded. "Continuing on, let's look at operations, which is about establishing the foundations to actually build, deploy, and sustain a differentiated advantage as a result of AI. For instance, you need to understand your data assets so you know what you can do with them, or what data you'll need to start collecting. You need to be aware that subtle differences in operations create subtle differences in your data, which can lead to exaggerated differences in AI behavior—in other words, data imperfectly reflects reality. You also need to understand how work is currently done so that you can know how to either augment or change it. You need a way to measure impact. And you need a plan for change management."

She leaned forward, as if she were about to share a sensitive piece of information. "This is where most AI initiatives actually fail. Sometimes, clients will come to us with an idea for how they want to use AI. But when we dig in, we discover they don't have the data necessary to make the solution work. Or they do, but it's siloed across different systems. Or it's messy and therefore unusable."

"And sometimes," Oliver explained, "clients don't understand their current workflows. They'll say 'We want AI to optimize our procurement process' but can't tell you how procurement actually works today. Who makes decisions? Based on what criteria? Where are the bottlenecks? All of that is critical information to build an AI solution that is actually useful."

"Plus," Ava added, "we've seen companies roll out AI solutions that end up being underutilized because they didn't invest in preparing their workforce. It isn't uncommon for employees to see AI—particularly when it's positioned as a way to fully automate a task—not as a tool but as a threat."

Mark shifted in his seat, crossing his arms. "But couldn't we cut head count if we deploy the right AI solutions?" he asked.

"We hear that idea a lot," Oliver said. "And in some cases, the answer is yes. But in most cases, using AI to cut jobs is shortsighted and can also seriously backfire. Here are just three examples of how: You can lose valuable institutional capital, you can reduce workforce flexibility, and you can alienate the very workforce you're encouraging to adopt these tools and that can lead to falling behind in AI."

Mark frowned.

"Besides," Ava said, picking up where Oliver left off, "sometimes employees don't trust AI for other reasons. If it's given wrong answers during development and testing, or if they don't understand how it works, why would they rely on it? Or if they simply don't know how to use it."

Julie nodded. "We spent two months on change management for our contract system. We created usage guidelines, did training sessions with Ava and her team, and tried to address any concerns we heard from employees. Without that, I doubt Anita and her team would have adopted the system so adeptly."

"So operations isn't glamorous," Ava concluded. "But without it, your AI strategy is likely to produce, well, nothing."

"We don't have time for all that hand-holding," Mark grumbled.

"Let's hear them out," Julie said firmly, shutting Mark down.

"Engineering," Oliver said, moving on, "is about having the technical capabilities to actually execute. That can either mean building solutions internally that create defensible advantages, or the ability to technically vet partners who can."

"At first," Julie said, "we didn't have the ability to do any of this. But Guidebridge helped us get started and have since helped us figure out the skills we needed to hire to maintain the systems they've built for us in-house."

My mind was swimming with all the implications. What was the last thing Julie said? I tried to focus.

"Can I get a copy of everything we talk about today?" I asked.

"Absolutely," Oliver said.

"So, as I was saying," Ava continued. "The key is how these three pieces work together. Strategy without operations is just talk. Operations without engineering gets you nowhere. And engineering without a strategy has you likely building things that don't move the needle. Conversely, strategy plus operations gives you an AI Center of Excellence. Operations plus engineering enables Machine Learning Operations (MLOps) and scalability."

"Can you elaborate on strategy plus engineering?" I asked.

"It's really the key. How do you know what to focus on and build without deeply understanding the overall corporate strategy? And how do you come up with an AI strategy that is differentiated and leverages the most cutting-edge technology if you don't have a deep understanding of what the state-of-the-art of the technology is?" Ava asked.

"It enables what every company is looking for: substantial and differentiated business value," Julie added.

Mark crossed his arms. "Sounds nice on paper."

"It's more than that," Julie said. "This is how we came to build our route optimization system. We started with a clear understanding of the most valuable business lever to pull. That may sound easy, but it took time, and I really feel like it was critical.

From there, we experimented with the latest open source models with Guidebridge's help until we understood which one would achieve the best business results, deeply understood our data, built a bespoke model based on our proprietary data, and put it into production while taking into account the change management needed for it to succeed."

"It's impressive how thorough and intentional you've been. How long did that take?" I asked, already dreading the answer.

"About a year, altogether," Julie said.

I felt my stomach drop.

"Clearly this isn't going to work," Mark said. "The board wants results now."

"Have you ever heard of Amara's Law?" Oliver asked, his tone professorial.

"What's that?" Mark asked warily.

"It's the observation that we tend to overestimate technology's impact in the short run and underestimate it in the long run. Everyone thinks AI will revolutionize everything by next quarter. It won't. But in ten years?" Oliver spread his hands. "The world will likely be more different than we can imagine."

"So?"

"So this gap—between expectations and reality—creates pressure for AI theater instead of actual transformation. Quick wins that look good in presentations but don't deliver results. We see it a lot. Building significant business value takes intentionality and time."

"I'm not trying to be critical, but that's easy for a consultant to say," Mark shot back. "But we don't just create slides for a living. Right, Sarah?"

Everyone looked at me.

Oliver shook his head. "I know that's the reputation of consulting, and well deserved in many cases. But, respectfully, moving fast without direction isn't progress. Here's the thing—you don't have to start from zero." He was addressing me directly now. "Julie's built something real with measurable success and impact. You could use it as a blueprint for the company-wide strategy. Show the board what's possible, even if it takes time."

"What? Just copy and paste?" Mark sounded incredulous. "That's ridiculous. We need to move quickly to show our clients and competitors that we're AI leaders." He turned to me, and I knew what was coming next by the zealous look in his eyes: "I think we should rebrand to FreightCo.ai."

Oh God. Not now, I thought.

"Interesting idea," Oliver said, with seasoned neutrality. "But why do you think that would help? Wouldn't that be creating the same dynamic of inflated expectations I just described?"

It was one of those neutral sounding questions—offered up as if on a silver platter—that didn't so much condemn the receiver in tone but left him with absolutely nowhere to turn. Mark, trapped as he was, became defensive.

"Well, you tell me, huh? You've got '.ai' on the end of your company's name. Why did you choose Guidebridge.ai?"

"Because," Ava said calmly, "AI is all we do. We're what I call 'AI native'—we started solely focused on AI, and AI professional services are one hundred percent of what we do."

Nobody needed to ask the next, obvious question: ***What percentage of FreightCo's revenue comes from AI, Mark?***

I needed to shift the conversation.

"But you're a services firm," I said quickly. "How should we think about AI's impact on our revenue as a logistics company?"

"That's a great question," Oliver said, clearly grateful for the off-ramp. "AI ROI for enterprises is complex. We'll dive into that—"

Mark's phone buzzed on the table. He glanced at it. His tightly composed expression went slack, and it seemed like the color drained out of his face.

"I need to make a call," he said, voice quiet, almost hollow. "Family emergency."

Mark gathered his things and was out the door before anyone could respond.

I watched him go. Part of me thought Mark was saving face, unwilling to admit he was out of line. The timing seemed too convenient: cornered in an argument, then suddenly a family emergency? But another part of me was concerned for him. He had seemed legitimately shaken when he left. Whatever was going on, I knew I needed to talk to him.

Chapter 10

Dinner and Drinks

The wine was good: bright and fruity, with a smooth finish. I swirled it in my glass somewhat absentmindedly, grateful for the chance to pause and reflect.

Across the table, Julie was laughing at something Oliver had said that I hadn't fully heard. The restaurant was dimly lit and hummed with quiet conversations that gave it a particular coziness. It was the collective sound of lives unfolding over good food and good company. This was a quality I sought out when I was on the road and missing my family.

"Should we wait for Mark?" Ava asked, gesturing to the empty chair.

I checked my phone. I had texted him earlier. No response. "Let's give him a few more minutes."

Part of me was worried—what if it really was a family emergency? But mostly, I felt relieved that he wasn't here. I sank deeper into my seat, which was surprisingly comfortable.

"You know," Julie said, "Sarah asked me earlier how I chose Guidebridge as our consulting partner. I never did tell that story."

"By all means," Oliver said, topping off her wineglass, "please regale us."

Julie took a sip of wine, her expression shifting. "Six months before I began working with Guidebridge, I hired one of the Big Four consulting firms."

"Which one?" I asked.

"Does it matter? I've come to believe that they're all the same when it comes to AI." She shook her head. "Two partners flew in for the kickoff. Promised us 'transformational AI capabilities' and 'market-leading solutions.' Once I signed, they handed us off to a team of junior consultants who'd clearly just finished a GenAI training module. They spent four weeks evaluating what we were already doing, then presented us with a deck of unactionable recommendations that was obviously recycled from previous clients."

Having experienced the frustration of overpaying for mostly useless slide decks from name-brand consulting firms myself, I asked, "How much did that cost?"

"I'm a little embarrassed to tell you, but let's just say it was hundreds of thousands of dollars for phase one. For that, I was told we needed to 'embrace digital transformation' and 'leverage data assets.' After that, I had no interest in phase two and fired them. I was polite about it, but firm. Then I spent a month being very, *very* methodical about finding the right partner."

"I developed a framework of the characteristics my ideal partner should have. I have it right here." Julie pulled out her phone and scrolled. "All the items I came up with fall into one of two categories: maximizing opportunities and minimizing risks."

"Can I see?" I asked, leaning forward.

Julie turned her phone screen toward me:

Characteristics of an Ideal AI Professional Services Partner

They help **maximize outcomes** because they:	They help **minimize risks** because they:
· Are focused on creating significant and differentiated value. · Have deep capabilities across the entire AI toolbox. · Help their clients transform into AI-centric organizations. · Meet clients where they are. · Take care to build a deep understanding of their clients' businesses.	· Are truly AI-focused. · Can help with strategy, engineering, and organizational transformation. · Are investing in keeping up with the technology to stay on the leading edge. · Understand the need to be responsible with AI.

The text was small, and I asked her to walk me through the framework.

"Sure," she said. "In terms of maximizing outcomes, the big firm talked a big game, but they weren't focused on creating significant value for us. They wanted to sell us generic platitudes. I also interviewed newer firms that were clearly just riding the ChatGPT wave. All they talked about was LLMs, but they didn't convince me how they'd create significant and differentiated value for us in particular. Guidebridge was different." Julie glanced at Oliver and Ava. "They came in asking about our specific business challenges and sharing how they thought AI could create real value for us."

She paused, then continued: "In a similar vein, the ideal partner had to have deep capabilities across the entire AI toolbox. Not just PowerPoints or the ability to hook up APIs to LLMs, but real, diverse technical depth. Can you build production systems? How many times have you done it before? The big firm had an 'AI practice,' but when I dug deeper, it was mostly rebranded analytics work with a veneer of generative AI implementation."

"That's a huge difference," I acknowledged.

"Exactly. But I also didn't want to become dependent on a consulting firm. I needed a partner who could help us build our own capabilities."

"That's rare," I said. "Most consultants want to stay forever."

"Exactly. But when I talked to Guidebridge, they were different. One of the first things Ava said was, 'Our goal is to transform you into an AI-centric organization. And if we do that, we'll make ourselves unnecessary.'"

Ava smiled. "That's still the goal."

"They also met us where we were," Julie added. "Our data wasn't in very good shape, and our team wasn't really ready. We had some ideas for use cases, but we didn't know if they could be built or if they were the right things to start with. Plus, we didn't want to fail right out of the gate and have people think we couldn't do AI. I expressed this, but the big firm kept talking vaguely about 'multidimensional approaches' and 'maturity models.' Basically telling us we weren't ready for AI without addressing my specific concerns. Guidebridge said, 'Show us what you have. We'll start there.'"

"Most companies are where you were," Oliver pointed out.

"And they took the time to understand our business," Julie said. "Not just logistics in general, but *our* specific operations, our constraints, our opportunities. That deep understanding is what made the difference."

"So you've worked with other logistics companies?" I assumed.

"No," Oliver said. "And we let Julie know that right up front."

"And I appreciated that," Julie said. "I felt like I could trust you guys."

I felt the same way.

"So that's the first category. The second category I was looking for was a partner who could help us minimize our risks. My first question was: Are they actually AI-focused? As in, not a consulting firm that added an AI practice last year, but people who really know this stuff. Are they experts at the top of the field?"

Oliver added, bashfully: "We're AI-native. Founded specifically to focus on AI, five years before LLMs hit the headlines. That singular focus has helped us attract and retain real AI experts."

"Right," Julie said. "Experts who could help with both strategy *and* engineering. Some firms are all strategy, no execution. Others can build things but not help figure out *what* to build. I needed someone who could help us figure out what to build *and* actually build it."

"On top of that," she continued, "AI moves so fast that if you aren't consistently learning what's new, your advice isn't all that relevant. I asked every firm: 'How do you stay current? What did you learn last month?'"

"We dedicate twenty percent of our time to learning and research," Ava said. "It's the only way to stay relevant. The field of

AI is moving too fast to expect to stay current without a significant investment."

I took a mental note to consider something similar for FreightCo.

"Last thing—and I can't stress this enough—they understood the need to be responsible with AI," Julie concluded. "Not just 'move fast and break things' like you hear from Silicon Valley, but thoughtful about the potential for bias in data, about ethics, about building systems that we could trust and our customers could trust."

"We've turned down clients with projects that didn't meet our ethical standards," Oliver added quietly.

"See?" Julie said to me. "Well, anyway, I interviewed about a dozen potential partners, and Guidebridge was the cream of the crop. After working together, I have to say I feel totally vindicated by this decision."

"Cheers to that." I lifted my glass, and we toasted.

"So, how long have you two been doing this?" I asked, after taking a sip of wine.

"Before I answer that question, let me back up a little bit," Oliver said. "Guidebridge came out of my experience of living through the start of the internet era. The commercial internet started in earnest in 1996, and, while it was hard to really tell what was going on at the time, in retrospect, the internet clearly changed the basis of competition forever. Internet-native companies toppled some of the most established firms of their time. Today, we all know the names of the tech giants that won."

He paused, and I noticed I was leaning ever so slightly toward him, wanting to hear more. Oliver spoke with a certain relaxed intellectual playfulness that made me interested in what he had to say.

"I started thinking about Guidebridge more than five years before large language models hit the scene. This was 2017. I was surveying the tech landscape, and I had this intuition that AI was going to represent the next big wave. It felt like the early internet in that there was speculation around it, and there were some truly impressive AI systems in existence—Deep Blue beating a chess grandmaster, for instance—but very little serious discourse about how it might fundamentally reshape business. *And*, at the time, it was almost impossible to find any professional services firms talking about AI."

He took a moment, gathering his thoughts. "You know, today, the big consulting firms say that AI represents the largest technical transition any of us will live through in our lifetimes. Back then, it wasn't so obvious. Interest in AI had waxed and waned going back to the 1950s. So starting Guidebridge back then was definitely a gamble. But as I was thinking about what organizations might need in this new era, I realized that, unlike in the internet era where software engineers had a relatively easy time picking up what it took to build websites and beyond, at the cutting-edge, AI requires skill sets that are pretty rare today."

He paused, taking a sip of water. "It's hard for software engineers to learn techniques from deep learning because the domains are so different. I'm not talking about hooking up LLM or other AI APIs. That's pretty straightforward. I'm talking about building cutting-edge AI models based on proprietary data, which was something I felt was necessary to help organizations build highly differentiated value. Does that make sense?"

"Yes," I said.

"Good. To do that, I realized I needed to build a firm that combined two things: this rare engineering skill set and what people traditionally think of as management consulting or strategy services, but exclusively focused on AI, not all the things people think about when they hear those terms."

"I joined to establish and lead the Strategy team," Ava chimed in.

"And I am sure glad you did," he said. "Because it was this mash-up of disciplines, strategy and engineering, focused on delivering differentiated business value, that was the initial guiding vision of the company. We put together a stellar cofounding team, and we've stuck to our vision ever since. For the kinds of engineering work that's more mainstream and less differentiated, we partner with other firms we trust who do that work. It's been incredibly gratifying. We've been able to attract and retain top talent, do meaningful work, and help our clients compete in this new age, all while helping them learn to do it themselves and responsibly."

"Wow. That's an impressive origin story," I acknowledged. "You were really ahead of your time. I'm glad Julie found you."

"Not to belabor the point," Julie said, "but that really was not easy. There's a lot of noise out there. In other words, lots of AI consulting firms tooting their own horns."

"Pardon me," our waiter said, arriving with our food. During the momentary break in conversation, I looked around and noticed, a few tables away, a man dining by himself, bent over and focused on his plate. That had been me on many occasions when I'd been on the road, and that recognition only deepened my appreciation for the opportunity to spend time with Julie, Oliver, and Ava.

"Sarah, how'd you get your start at FreightCo?" Ava asked.

"I'm relatively new to FreightCo," I said, smoothing my napkin in my lap. "I started in marketing out of college and became a brand strategist and eventually the CMO of a global retailer based in Boston. FreightCo recruited me as CEO two years ago—they wanted someone who understood the customer side of logistics. We—my husband, David, and I—moved the family from Boston to Atlanta."

"That must've been a challenge," Oliver acknowledged. "How many kids do you have?"

"Three. Emily, my oldest, is sixteen. Jake is thirteen, and our youngest, Lily, is eleven. And yes, it's been a transition . . ." I lingered on the sentiment, and then added: "It's been hard, yeah. David had to put his policy consulting practice on hold. The kids had to leave their friends. And now, well, you've probably heard. On Monday, the board gave me ninety days to come up with an AI strategy or they'd find someone who could. On Wednesday, after Apex's announcement, they cut it to thirty days."

"I hope you know how ludicrous that deadline is." Oliver took a bite of food.

"I know how it feels to run on four hours of sleep for a week." This got a few laughs. "But seriously, I know it's a lot. I feel the pressure. Our competitors are promising AI. The board wants AI.

It's AI this and AI that. I'm not at my wit's end, but I'm getting there."

"You're not the only one," said Ava. And then, noting my surprise, she added casually, "Not in terms of the work we do. I love that. But the discourse around AI. The hype, the headlines, and the broad generalizations. Some of our clients have studied up and are practical and realistic. Others come to us thinking that we can just sprinkle AI fairy dust around and magic will happen. Some people treat AI like it's this panacea that can fix anything. That's just not true. AI can be incredibly powerful, but you need to know what problem you're trying to solve and what technique you're going to use to solve it."

"Like we said earlier, AI isn't a tool. It's a toolbox," Oliver added. "And most people don't see it because they're fixated on the newest and shiniest tool, generative AI, which is also the least mature from a risk point of view, given things like hallucinations."

It occurred to me that, for Mark, generative AI was a hammer, and everything looked like a nail.

Ava continued: "Saying we need to invest in AI is like saying we need to invest in transportation. It's a start. But what kind of transportation? Cars, boats, planes? When most people today say 'AI,' they're referring to generative AI or chatbots. Or they just say 'GenAI' outright. But that's like calling every vehicle a 'car' just because that's what's in the ads."

"What's in the toolbox?" I asked.

"Many things," Oliver said. "Here, I actually have a printout of a slide we use in my bag for a presentation I'm preparing for."

He reached into his backpack, removed a folder, and passed a piece of paper to me.

The Full AI Toolbox | Part 1

Analytical AI

Vision & Documents
Image: Object Detection; Classification, Change Detection, Segmentation, Facial Recognition; Pose Estimation

Video: Action Recognition, Video Understanding; Tracking

Documents: OCR, Layout/Table/Field Extraction, Handwriting Recognition, VQA

3D: Depth Estimation, Structure from Motion; 3D Object Detection; Volumetric Segmentation

Language & Code
Text Understanding: Classification, NER, Part-of-Speech Tagging, Topic Modeling, Sentiment/Intent Modeling

Retrieval: Embedding Models, Vector Similarity, Approximate Nearest Neighbor, Hybrid Lexical & Semantic Search, Re-Ranking

Speech & Audio
Automatic Speech Recognition; Speaker Recognition; Emotion Recognition; Diarization

Audio Event Detection; Source Separation; Sound Localization

TTS, Voice Conversion; Voice Cloning

Tabular & Time-Series
Classification, Regression, Forecasting, Anomaly Detection; Imputation; A/B Testing

Recommender Systems

Decision Science: Survival Analysis, Uplift Analysis, Causal Inference

Graphs
Node/Edge/Graph Classification

Knowledge Graph Construction

Community Detection; Graph Anomaly Detection

Generative AI

Language and Multi-Modal: LLMs for Text Generation, Vision-Language Models (VLMs)

Code Generation & Review

Retrieval-Augmented Generation (RAG), Context Engineering

Agentic Systems (functions, APIs/MCPs, browsers, CLI, planners, verifiers)

Image/Video: Diffusion Models for Image and Video Generation (T2I, I2I, T2V, I2V, etc.), Image Captioning, GANs, VAE, Navigational World Models (action-conditioned video generation)

Audio: Audio Generation & Editing (speech & music), Speech-to-Speech

3D: NeRF, 3D Gaussian Splatting (3DGS)

Math: Theorem Proving

Biology: Protein Design, Structure and Interaction Prediction

Chemistry: Molecule and Material Design

Physics: Fast Physics Surrogates for PDE Solvers

The Full AI Toolbox | Part 2

Physical AI

Digital Twins

SLAM, Navigation

Vision-Language-Action Models (VLA)

SIM-to-Real

General Algorithms and Best Practices

Responsible AI
Alignment: RLFH, Constitutional AI, Etc.

Red-Teaming: Jailbreaks, Prompt Injection, Bias/Safety Stress Tests

Transparency: Model Card, Data Sheet, Feature Attribution, Mechanistic Interpretability

Privacy: Federated Learning, Differential Privacy

Model Architectures
ConvNet, LSTMs

Transformers, State-Space Models (SSMs)

Gradient Boosted Trees

Kernel Methods

Generalized Linear Models

Training Algorithms
Pretraining, Supervised Fine-Tuning (SFT)/Instruction Tuning, RL with Human Feedback (RLHF), RL with Verifiers

LoRA, Model Distillation

Self-Supervised/Contrastive Learning

Inference Optimization
Model Quantization

Speculative Decoding, KV Caching

Unsupervised & Semi-Supervised Learning
Clustering

Dimensionality Reduction (PCA/UMAP/t-SNE)

Active Learning

Model Deployment/Productionalization
MLOps: Experiment Tracking, Shadow Deployment, A/B Testing, Deployment, Performance Monitoring

Orchestration: Model Context Protocol

Evaluation and Benchmarking: Model Evaluation, Human-in-the-Loop

Containerizing: APIs, Etc.

Looking over it, I couldn't help but notice how many of the things listed were things I didn't understand.

Perhaps it showed on my face or in my apparent silence, because Oliver said: "If I could just say, Sarah, you don't need to be an expert in these things to meet the expectations of your board. We can help you come up with a strategy, a clear plan, that leverages the right AI tools for the right job, and from there, you can learn and adjust as you go."

"*And*," Ava added, "you should know that it's often the case that AI solutions that deliver value in production leverage combinations of AI tools. And pretty much every executive we work with feels overwhelmed, at least to some extent. You're certainly not alone in that."

I thanked them, and we talked at length about a variety of things, from our children to hiking to restaurants in the Austin area. I promised to take them out whenever they were next in Atlanta. Eventually, as the evening was dying down, the waiter appeared with the check. Oliver reached for it, but I brushed aside his offer.

"Our treat," I said. "Thank you for everything."

Outside the restaurant, we said our goodbyes. Julie, Oliver, and Ava were all headed home, albeit in separate directions. I elected to walk the short distance to my hotel. It was a warm and inviting spring evening, just after dusk. There was still enough light to see by, and I felt safe and unburdened walking. I knew I would crash from lack of sleep soon, but for now, I was awake and alert and in tune with some inner part of me that felt cautiously optimistic we would figure this out. And, if we did not, I would have at least done all I could. I had a strong sense we had the right partners, and I had the right rising leader in Julie. Mark, for all his flaws, could be managed. And if he couldn't, we'd need to part ways.

As I walked, it dawned on me that my hotel was farther away than I had realized. I decided to call David and see how he was doing.

"Hey," he answered. "How's Austin?" Hearing his voice was comforting, and after two decades of marriage, it was still one of my favorite sounds. Especially when I was on the road.

"It's good!" I said, a bit winded as I walked uphill. "I'm learning a lot. I think I'm starting to get what it is we're dealing with. Like, the technical details are still opaque to me, but I get the shape of it, you know? I think we've been trying to buy a concept when really we're dealing with a broad collection of specific tools that need to be learned, assessed, applied, maybe even discarded."

"That's heady stuff," David said. "I'm glad you're having a good time."

In reflecting on the string of words I had just rattled off, I realized that I was, perhaps, feeling that second glass of wine. "How are things at home?" I asked.

David told me about his workday and dinner with the kids (they'd gotten takeout) and more as I got back to the hotel and then, with a not insignificant amount of bodily relief, my room.

Sitting on the edge of the bed, I took off my shoes. Then I lay down, stretching my arms out wide, and talked to David for a few more minutes with my phone lying on the pillow next to me. Eventually, I told him I was getting tired and was going to bed.

"I love you," I said.

"I love you too," he said. "See you tomorrow."

We hung up. I yawned, and looking down at my phone once more, noticed I'd missed a text from Mark.

Sarah, I know it's late but could we talk? I'm in the hotel bar if you're still up. I need to explain about today.

I scrutinized the text. He'd sent it about a half hour ago. I was tired in my bones, and not unimportantly, I was relaxed from the evening and by this point thoroughly out of "work mode." But something in Mark's phrasing—*I need to explain*—nagged at me. It tugged at the part of me who considered herself a leader who showed up to help people when it mattered most.

Solemnly, and with some hesitancy, I slipped my shoes back on, grabbed my room key, and headed for the lobby.

Chapter 11

The Hotel Bar

There he was, hunched over a glass of whiskey, shirt rumpled and hair tousled by the persistent movement of his hand scratching at his temple.

"Mark," I said, approaching cautiously.

He looked up, attempting a smile. "Sarah. Thanks for coming down. Sorry I had to rush out of the meeting today."

"It happens," I said. The bartender offered me a drink that I politely declined. "But I'm worried about you. What's going on?"

"Nothing. It's all fine." He took another sip, studying the liquid in his glass.

I sat down next to him. "Mark, you reached out to me. What's on your mind?"

"No, I'm sorry I bothered you." He waved dismissively. "I know you must be tired."

"Mark." I kept my voice steady. "Obviously something's bothering you. I know you got a phone call from home. Is everything okay?"

"Everything's fine," he repeated, gripping his glass tightly, as if for support.

I tried a different approach. "You know, I see a big difference between how you're approaching AI and how Julie's approaching it. I think we both have an opportunity to learn from her."

"Yeah. Sure." The words came out flat, automatic.

"Okay, stop. We need to talk about your division and AI—not necessarily now, but soon. And as for whatever's happening at home . . ." I paused, choosing my words carefully. "Look, work is important, but that's the stuff that really matters."

He finally met my gaze.

"I don't know if you know this," I continued, "but when David and I first moved to Atlanta two years ago, it was hard on our marriage. We uprooted everything for this job. It was hard for him. For the kids. For me. I felt a ton of pressure with this role, and I felt guilty for my family's unhappiness. It's taken a lot to get to where we are now. Therapy has helped."

I took a breath. "David and I are in a good place now. But it took real work."

Mark studied me with silent apprehension—as if weighing up whether to share with me what was on his mind. Gradually, and then all at once, he capitulated.

"I have no idea what to do at home," he said quietly. "My kids are struggling. My wife is talking about a divorce. I come to work and try to forget about everything. I'm throwing myself into my job, but I can't concentrate. I feel like a fraud here, and everything's falling apart at home."

I paused to absorb what he'd said. "Mark, I'm so sorry to hear all of this—I really am. And I really appreciate you being so vulnerable with me and letting me know. You've got a lot on your shoulders. But let me ask you this—if you are okay with it. Do you think your marriage is really over or do you think it's possible to save it?"

He hung his head. "I don't think my wife wants a divorce, but she says she's unhappy. She says I'm distant and she doesn't feel like we're connected. I would be devastated if she left, but I don't really know what she's trying to say or what to do about it."

I considered this for a moment. "It's good that your wife is open to saving the marriage, but it sounds like it is hanging in the balance. If you're okay with it, I've got some unsolicited advice. The first thing I'd recommend is to work on your own stuff. Really try to understand what your wife means and how you're contributing to what she's feeling—or not feeling. Do you have a therapist?"

"Yeah, but . . ." He hesitated. "I only saw him a few times. I didn't find it helpful."

"Did you hold anything back? Tell him the whole truth about what's going on?"

During the long pause Mark fussed with his glass and continued to avoid my gaze.

"No. If I'm being honest, not really."

"That's where you start," I said gently. "Go back. Tell them everything. The work stuff, the home stuff, what your wife is saying, how you think they might be connected. All of it. I'd encourage you to be really vulnerable. It's hard and takes a lot of courage, but the therapy I've been doing and the couples work David and I have been doing has helped me understand it's foundational for a happy marriage, for success at work, and, really, in life."

I paused for a moment. "Mark, family matters more than any of what we're working on. Take some time off. When you come back, I'm going to need you to bring some of that vulnerability with you. But we can talk about that then."

He pushed himself up from the bar. "I should book a flight home."

"Good idea."

As he began to walk away, he paused. Stone-faced and a bit gray, he nodded—a solemn, lonely gesture.

"Thank you."

"We all need help sometimes, Mark. Asking for help is a sign of strength. I'll be thinking about you. Good luck with everything."

Chapter 12

Wayfinding

Ross, Julie's CTO, shook my hand. "It's great to meet you in person, Sarah."

"You, too, Ross. I'm looking forward to learning more about your approach to AI."

We were meeting in the same conference room as yesterday. Julie had texted me earlier in the morning to ask me for my coffee order. Now, as she walked in balancing a carrier with four cups in one hand and another with two cups in her other hand, I helped her distribute the drinks.

"Thank you, Julie," Oliver said, taking his.

Julie nodded.

"Where's Mark?" she asked, gesturing to the drink she'd brought for him. "I didn't know what he wanted, so I just got him a coffee."

"Mark won't be joining us today," I explained. "He has to attend to a family emergency."

"Oh." Julie sounded a bit startled. "I hope everything's okay."

"I'm not exactly sure what's going on, but I hope so too," I said.

After a few minutes of checking in on everyone, I suggested to Julie that we should get going. She deferred to Ross.

"Your show to run."

"Okay. Let's do it." Ross ambled over to the chair across from me and sat down. He had a deliberate yet relaxed way of carrying himself. "I thought we could share our approach to our routing solution with you, but hopefully cover a few more technical points along the way. How's that sound?"

"Sounds good."

"Okay, perfect." He rolled up his sleeves and, briefly, deliberated about how to best tell his story. "So, when we first started working on our routing solution, we thought we were taking on a simple optimization problem: find the most efficient routes such that we could make rapid, on-time deliveries. We had years of route data, so we thought we'd train a model to find the most efficient path from A to B. That didn't work. For one thing, 'efficient' is a broad term with multiple underlying concepts: speed, cost, and distance, to name a few. And even when we defined it with clear terms, it didn't solve our *actual* business problem."

"Right." I nodded. "I heard about this earlier this week. Nine rapid deliveries don't matter if the tenth doesn't arrive at all."

"Exactly," Ross said. "The AI found incredibly efficient routes. But what you won't get with just optimizing for fastest routes is robustness. If a highway was cut off or if some country road was flooded or whatever the case may be, those 'perfect' routes can always fail."

"How'd you fix the problem?" I asked.

"The short answer is that we aligned on a clear business goal. But there was a lot involved in that. I think it may be helpful if I elaborate on what's happening 'under the hood,' so to speak, and why aligning on a clear business goal is so important." Now that he had gotten going, he spoke with the intensity of someone who spent their days solving problems that mattered.

"When we're training any AI solution using deep learning, the model is looking at the data available to find the way to mimic the results captured in that data—the relationship from any given piece of input to its corresponding piece of output—and make the fewest mistakes it can based on the data it was given."

He paused, considering his words. "Think about it as a sort of a 'win at all costs' mentality. With AI, there's no real ideas of sportsmanship or artfulness. It just needs to win."

"Win? What's that mean?"

"Winning in this case means getting the right answer. Making the fewest mistakes possible based on the data it was given, even if that means 'cheating,'" which he put in quotes with his fingers. "Which is a problem when working with deep learning models, because of that 'win at all costs' approach. Not to mention that it's difficult to spot. For instance, with a person, we'd recognize certain shortcuts to winning as problematic: like cutting across the infield in a track race or picking up the soccer ball and carrying it into the goal. Plus, we have social pressures and a sense of ethics. We tend not to do things like that. But AI doesn't inherently have a set of values. It has goals, data, and feedback, which often means it finds its own kinds of shortcuts in the data.

That might mean leaning into bias that's already present in the data to make predictions that are *technically* accurate within the context of the biased dataset but not correct or aligned with what someone might actually want."

Oliver chimed in. "Another way to think about this is that if your data accurately captures the input and related output that form your business problem, then AI can probably 'win' or solve your business problem with few or no unintended consequences. But if your data is not quite what you want or has a lot of other things muddled up in there, then an AI model doesn't have any inherent sense of which patterns it should or shouldn't learn. It only knows which patterns get it to that 'win' state as fast as possible. And it learns those."

"Despite the consequences," I said. The idea that you could train an AI to be accurate but totally wrong both fascinated and worried me.

"Exactly."

"So how do you make sure it learns the right patterns?" I asked.

"That depends entirely on your business problem," Ross said. "It's a really hard thing to do and it can't so much be predicted as discovered. That's where this process of Wayfinding that Guidebridge taught us comes in."

Ava gestured toward the projector, and presented the following slide:

Wayfinding

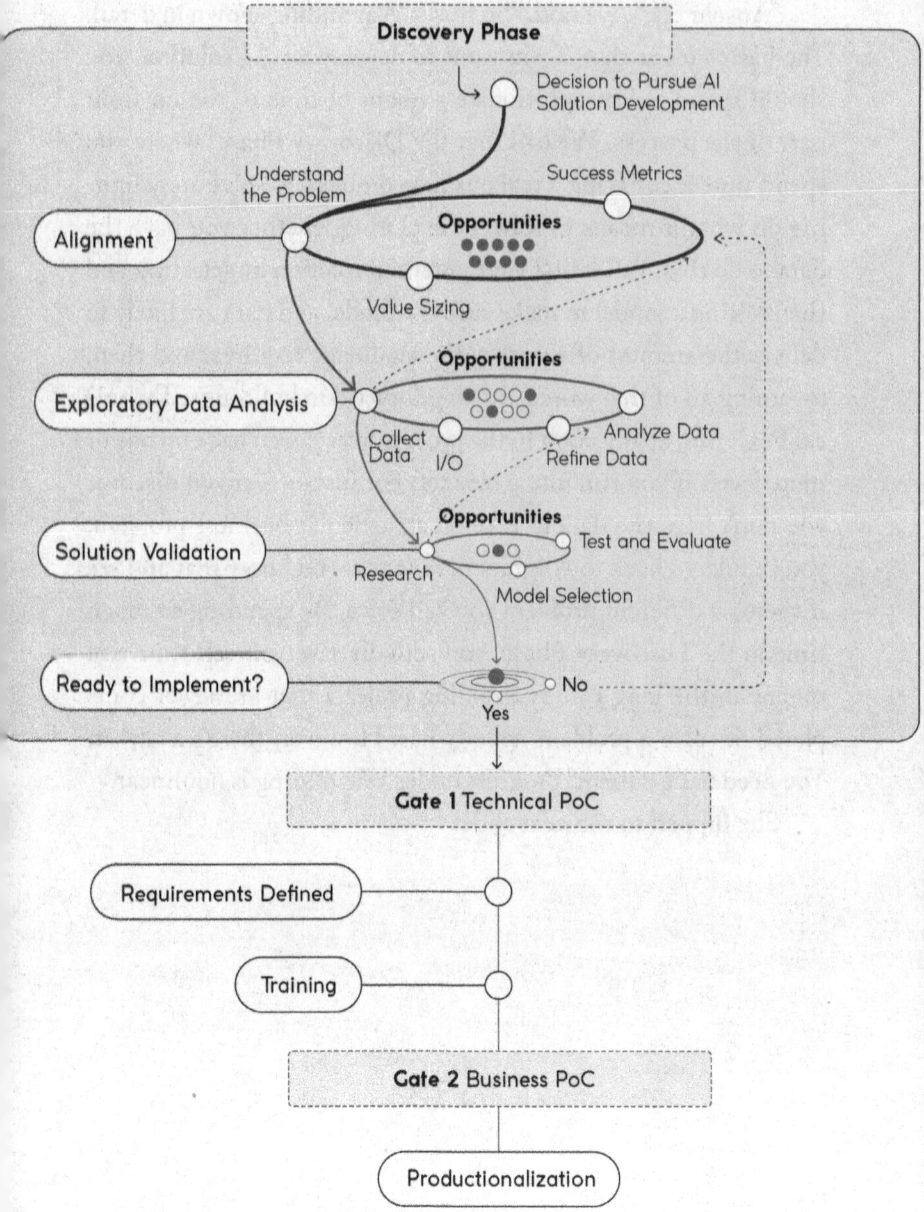

Discovery Phase
Decision to Pursue AI Solution Development
Understand the Problem
Success Metrics
Alignment
Opportunities
Value Sizing
Opportunities
Exploratory Data Analysis
Collect Data
I/O
Refine Data
Analyze Data
Opportunities
Solution Validation
Test and Evaluate
Research
Model Selection
Ready to Implement?
No
Yes
Gate 1 Technical PoC
Requirements Defined
Training
Gate 2 Business PoC
Productionalization

"That's quite a diagram," I said. "Help me make sense of it."

"Absolutely," Ava said. "So this is Wayfinding shown in detail. The basic idea is that if you want to develop an AI solution, you should spend a disproportionate amount of time in the up-front part of the process. We call that the Discovery Phase, where you spend time identifying a real business problem to solve and aligning on what it means to solve it, exploring whether you have the data to do that, validating your potential solution by selecting and then testing a model to make sure the model and data are likely to deliver the amount of accuracy the challenge requires, and then, assuming all of that works, moving forward to building a Technical PoC. But, at any point in the process, you can go back up one or more levels if you run into a blocker. For instance, if you discover you don't have the data to take on a particular business problem, you should go back to Alignment now that you know that and see if there's a different problem you can solve. By spending so much time in the Discovery Phase, you actually save yourself time and money in the long run by avoiding projects that are never completed or solve a problem nobody has. However, there's a caveat: You need to be patient. Progress under Wayfinding is nonlinear."

She flipped to the next slide:

AI Development Progress Under Wayfinding

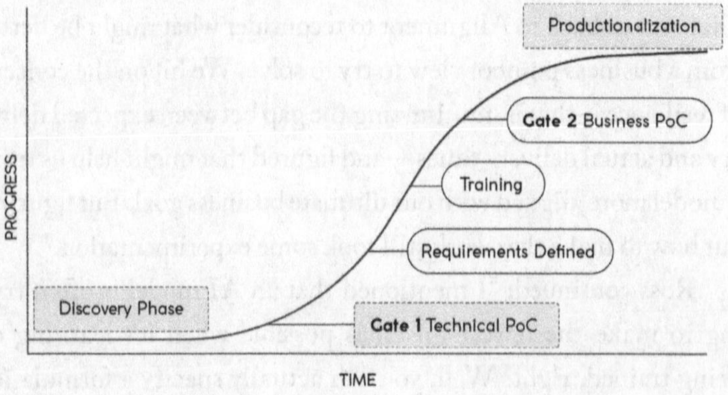

"Progress can feel painfully slow during the Discovery Phase. But once you get to the first gate, the Technical PoC, which, again, can take a long time, progress tends to happen quite quickly. If you have the problem clearly defined, the data available, and the model selected, actually training the model can happen quite rapidly—say, a matter of weeks. That's why you see the 'S-curve' above. Going from a Technical PoC to a Business PoC is a rapid process. But then, once you're at Productionalization, progress tends to slow down again as you hammer out any remaining kinks in the model, get it to the level of accuracy needed, or address edge cases to ensure it's deployment-ready."

"Okay . . ." I said. "I think I'm starting to get it. But I'd benefit from a real-world example. How did Wayfinding work in terms of developing your routing system?"

"I can take that," Ross said. "As I mentioned, we started out by thinking we were solving for the most efficient routes. That's what we agreed on initially in that top Alignment step. But during the next step, Exploratory Data Analysis, or 'EDA,' we realized our on-time delivery rates didn't correlate strongly with route distance. That sent us back to Alignment to reconsider what might be better from a business point of view to try to solve. We hit on the concept of resilience—that is, minimizing the gap between expected delivery and actual delivery times—and figured that might help us train a model more aligned with our ultimate business goal. But figuring out how to make that work still took some experimentation."

Ross continued: "I mentioned that an AI model is often trying to make the fewest mistakes possible when it's learning or being trained, right? Well, you can actually specify a formula for how bad any given mistake is and nudge the model toward a more nuanced goal, or as we say, objective set. One of the things that has become more popular recently is actually mixing together several different criteria. So you can think of it almost like a cooking recipe." He paused. "It's like one-part-delivery-speed to one-part-worst-case-outcome or something along those lines. You can kind of mix and match these attributes, and you can also add weights to them. For instance, you could say 'five-parts-speed, one-part-resilience' as one variant of calculating resilience. That's what we ended up doing with our model."

"It's important to note," Ava added, "that ideally, we nail the objective in the Alignment step, but it's not uncommon for us to have to iterate. Either way, we want to make sure we get it right before we progress past Solution Validation to the step of creating a technical proof-of-concept. At that point, we want to be assessing the ability of the solution to deliver on our stated objective—for instance, what percentage of the time is it 'right,' as opposed to figuring out what our objective should be."

"Got it. Don't build the wrong thing—or the right thing with the wrong goals. Check."

"In a nutshell, yes." Oliver smiled. "You've likely heard that a huge percentage of AI models don't make it into production. Wayfinding is the agile-based methodology for AI projects that Guidebridge developed to help make sure that doesn't happen—and it works!"

"But if I may take this a step further," Ava said, "if you do need to start over with a new objective, you might find that you don't actually have data to support that objective. If so, you want to know that as soon as possible. It's not a great thing to find out six months into a project when you've already burned through your political or actual investment capital. Or your team is exhausted. Basically, there are a lot of bad outcomes you'd want to avoid, none of which happened here, thankfully. But our machine learning engineers who worked on this solution with Ross and his team, they've seen it all before."

Noted, I thought. "How long did the whole development process take?"

"The short answer," Julie said, "is about a year. But collecting the data for it started the year before that. We started collecting data even before we hired Ross. That was back when I wasn't sure what we were going to do about AI, but I was learning enough to recognize data was important. We started tracking not just delivery times but driver feedback and customer satisfaction scores. And we found a vendor for weather and historical traffic data. We didn't know exactly how we'd use it all, but we knew enough to think we'd need it for something."

"Julie had good foresight," Oliver said. "Data isn't just a by-product of doing business anymore. With AI, it's an engine of doing business."

"Wayfinding ties into that, too, in a way," Ross said, "In EDA—sorry, Exploratory Data Analysis—you have to ask yourself: Do we have the materials on hand to solve this business problem or not? Luckily, we did."

Although I'm not sure we really do beyond Julie's division, since I didn't even know that was something important, I thought.

"What else can you tell me about the Solution Validation Phase?" I asked. "How does it actually work? And how can we make sure we don't create a model that 'wins' while creating unintended consequences by not really being aligned with what our business goals are?"

"Those are great questions," Ross said. "Let me try to explain what's going on in that phase. Imagine you have an AI model that's watched five hundred baseball games, and at the seventh inning stretch, it has to guess who's going to win the game. It has some idea, you know, because it's picked up on patterns. If you show it game five hundred and one up to the seventh inning, it can make a prediction about who's going to win that's probably pretty good. In the Solution Validation Phase, we'll assess how good the model is, and we'll also test it to see how it performs on imperfect data. For instance, imagine that in the context of your business problem, your model only has access to the radio broadcast of the game. Can it still predict the outcome? Probably. But it might struggle if it's picked up on visual cues—like the scoreboard—and overlooked auditory ones in training."

"We also recommend training only on data that the AI model will encounter in production to try to avoid that kind of problem," Ava said.

"Okay, thanks for that. What about going from proof of concept to production?" I thought of Mark's failed chatbot.

"I'm glad you asked. Most people think building an effective AI model is just about having a good idea. But ideas are plentiful. Taking them from idea to proof of concept, or 'PoC,' is hard. But going from PoC to production is even harder. It's where most projects die," Oliver said gravely. "As I mentioned, only a small percentage of AI projects ever make it from proof of concept to production. Often this is because they aren't suitable to perform under real-world conditions. With Wayfinding, our hit rate is much better, around ninety percent."

"That's really remarkable," I said.

"Let's talk about another aspect of Solution Validation in the Wayfinding model. Think about Google Maps," Ava said. "It's immensely helpful, but imagine if it needed three minutes to calculate each turn. Nobody would use it!"

"Fair enough," I said.

"The time it takes to respond is called 'latency,'" Ross explained. "In addition to that, you would also need to consider cost and scalability. For instance, if everyone uses your system, can you handle the computational workload? What's the cost of doing so?"

"Plus, you need to consider the risk of being wrong," Oliver said. "If you're Google Maps, and you're right ninety-nine times out of a hundred, but every hundredth time you send someone on a catastrophically wrong turn, that's not good and would seriously impact your business. Plus, one trip probably has a minimum of ten turns, so that ninety-nine percent is felt as ninety percent accuracy to the user."

"This is also why we tend to keep humans in the loop," Julie explained to me.

I nodded. They were throwing a lot at me, but a picture was beginning to take shape in my mind.

"In general, we find it helpful to define clear success criteria that 'gate' or checkpoint going from PoC to production," Ava said. "Defining what your definition of 'done' is, up front, including things like performance targets, latency requirements, integration tests, and adoption goals. Otherwise, you might get stuck in a cycle of 'almost there.'"

Ava gestured to the slide again, tracing the S-curve up with her finger.

"Traditional waterfall development tends to be linear," she explained. "You know the requirements, you build steadily, and progress is visible. Even traditional scrum expects relatively steady progress sprint by sprint. But AI is different. When you're building AI capabilities, progress is usually slow at first while you're discovering and aligning. Then progress is rapid once you understand the problem. Then it flattens again for the 'last mile' of refinement and deployment."

"Experienced agile practitioners might recognize this pattern," Oliver added. "It's why many teams invest in what's called a 'Sprint Zero' for discovery. Wayfinding makes this pattern explicit and extends it appropriately for AI projects."

"I initially underestimated how much time my team would

need to spend on this," Ross admitted. "If it wasn't for Julie clearing my schedule, pulling in the right people, and dedicating significant resources to this, we probably would have succeeded in building a model, but it wouldn't have delivered the impact we desired. In order to succeed, we needed a project champion. That was Julie. Still is," he added quickly.

Julie acknowledged the praise with the modest gesture of a raised hand. I studied her for a moment. She hadn't been vocal during the session. In fact, she had probably said the least of anyone in the room. And yet, everything I was learning was, in some way, the result of her leadership. I wondered why she wasn't seeking more praise for it. But at the same time, I was grateful to her for creating the space for me to learn. For bringing me into the fold when I needed allies the most.

I had spent a lot of time and emotional energy managing Mark. And it dawned on me as I looked at Julie sitting with her hands in her lap that I hadn't shown her the same attention. I made a mental note to find more time with her.

I checked my watch: 11:57 AM.

"Excuse me," I said. "I have to take a call at noon. This has been extremely helpful, and I really appreciate you taking the time."

"Of course," Ross said. "If you have any additional questions, I'd be more than happy to chat again."

I thanked him and headed out into the hallway. This time, I had scouted out a room in advance. Setting my things down, I dialed the phone.

Dante, my friend who ran a SaaS company, picked up after a few rings and, sounding upbeat, asked me how I was doing. I told him about the board's ultimatum, the pressure we were feeling from Apex, and my time in Austin.

"You're doing the right thing," he said. "Spending time in Austin. I had to do something similar. Not in Austin, but with one of my teams in Seattle."

"And?"

"And it's going well!" Dante was typically an upbeat guy, but he sounded especially chipper today. "We've always been pretty progressive as a company, and I felt like we had a good base to build on, both in terms of data and the institutional willingness to make changes. We had to hire additional people—machine learning engineers and data scientists—because the traditional software development skill set isn't the same as the skill set needed to build and maintain AI models. But we've gotten things going."

"It's funny you should mention hiring," I said. "I'm meeting with the Head of People for the Austin Division this afternoon. I expect she'll have a perspective on that as well."

"Yes, ask her. The AI talent market is tight, as I'm sure you know."

"Yes. So what are you doing? Can you share?"

"Somewhat, yes. We've deployed a variety of productivity tools internally—think AI coding assistants—and we've got a flagship AI product in development. We also have a product road map and a clear vision for where we want to go. I really believe we've built something valuable."

"Congratulations. That's really great."

We chatted a few minutes longer, and I asked him a few questions about what they were doing right. Then I wished him well and we hung up. Closing my eyes, I leaned against the wall and exhaled. I felt a pang of envy for Dante and how straightforward everything seemed with his company. But also, after spending more time with Julie's team and reflecting on what I'd learned, I could see that we were on a similar trajectory. It would take work, but I was gaining confidence we could get there.

Chapter 13

Everything to Do About People

The café overlooking Lake Austin was half-full at 2 PM. We'd found a table by the window. Julie had suggested the change of scenery after lunch, and I was grateful she had said something. While I was looking forward to meeting her Head of People, I needed a coffee.

"August should be here any minute," Julie said, checking her phone. Her fingers drummed lightly on the table.

"Everything all right?" I asked.

"Yes . . . she's just not usually late." Julie glanced toward the entrance, brow furrowed.

I nodded and checked my watch. My flight wasn't until 5:30 PM, but I was already looking forward to being home with David and the kids. I crossed my legs and shifted my suitcase that was next to my chair to ensure it wasn't in anyone's way. My neck ached, and I didn't expect the flight home to make it any better. It had been some week.

I thought about using the time to talk with Julie about her career goals, but I was worn out and August had been about to arrive for the past ten minutes. I suppressed a yawn.

"There she is," Julie said.

I turned to see a woman walking briskly across the café toward us, thermos in one hand and phone pressed to her ear in the other.

"Right, thank you. We can talk more later." She hung up, and, in a split-second, shifted into professional mode, before greeting me.

"So good to meet you in person, Sarah." Extending both arms toward me, August said: "I'm a hugger—is that all right?"

"Absolutely—I am too!" I stood and quickly embraced her, and we took our seats.

"You know, August, this morning I met with Ross. We spent about two hours going over the cutting-edge work your team is doing in AI. It's totally unlike anything else being done within FreightCo right now. From what I can tell, you've built a distinct, impressive culture and a team of high performers here in Austin. I would love to hear more about your approach."

Her expression flashed from excitement to a deliberate restraint to something like acceptance.

"That's kind of you to say, Sarah. But I want to be up front about something—we just lost one of those high performers."

I sat back. August's directness caught me off guard. The news hit me like a punch. Was it Apex? Again?

"Who?" Julie asked.

"Andy Bertrand," August explained. "I only just found out. Apex offered significantly more than his current salary."

"Shit," Julie said. "He's brilliant."

August took a sip from her thermos. "I suppose Apex knows our people are worth poaching." There was a slight edge to her voice, but I could see her composing herself in real time. "But, you know, you can't build a retention strategy based on keeping everyone. You build it on creating an environment where people choose to stay. Andy has a kid starting college next year and another in high school. I can't fault him for needing to take care of his family. But damn, it stings. He was one of our best."

"How many people have you lost this year to Apex?" I asked.

"This is our second," August said.

"Oh, only two? How do you compete with people paying more than we do?"

"We try to offer the best version of everything on top of salaries, which we target at the midpoint of the market for each of our roles," August said. "This is actually what I wanted to talk to you about. A lot of the work we have done to support our AI initiatives has involved strengthening our people processes and culture, because AI is a massive pressure test on them."

"That's interesting," I said. "Tell me more."

"Well, for starters, top talent will leave if your people processes aren't good enough. But more than that, AI fundamentally changes the employee experience. Think about large language models: They're incredibly accessible to employees. We discovered that pretty quickly. Many of our employees were using AI, but in secret. It created this elephant in the room that a lot of companies still haven't really addressed." She paused. "Julie, to her credit, suggested we tackle the topic head-on."

"And?" I asked.

Julie took a deep breath, hands folded in her lap. "Pretty soon after ChatGPT came out, I stood up in an all-hands and told everyone I didn't fully understand AI, but that it seemed important. I said that we were all going to figure this new technology out together and that I was going to do everything I could to learn alongside them. And that I'd probably ask a lot of stupid questions along the way. I also told them that we would be transparent about any potential impacts to their jobs as a result of AI. Our aim was not to replace anyone, but to build and use the technology to augment ourselves. All of us, including me."

"You could feel the change in the room in real time. There was this tight, nervous energy and Julie's speech just let the air out of the tire."

"Kudos to you," I said. "That took courage."

"No kidding! I worried it could destroy my credibility. But deep down, I think I knew people were waiting to see if I'd bring up the topic and be vulnerable, so in some ways it felt like a foregone conclusion. I couldn't *not*, if you see what I mean."

"We then built on that foundation of trust," August said, "and launched a pulse survey tool, which is something we'd been researching and demoing at the time. We wanted to make sure we were understanding employee sentiment. The software sends out pulse surveys every week, and managers can see the results in real time."

"Every week?"

"I know," Julie admitted. "It seemed excessive to me, too, when August first pitched me on the idea. But the surveys only take our employees a few clicks to complete, and I'm glad we did it. They have the ability to add comments, if they want, and it's all anonymous."

"When you survey frequently, patterns emerge quickly," August explained. "This is helpful in the traditional sense, but especially given the fact that AI is accelerating many trends, having more rapid team sentiment data means we can see what's going on and act on things we see in the data quickly."

"Can you give me an example?"

"Sure. After we hired our first machine learning engineers to work with our data scientists, we noticed in the survey results that they really wanted opportunities to upskill, to stay on the cutting edge of the field. So we tried giving them Friday afternoons for that."

"You're giving up ten percent of their time," I said, doing the quick math. "Has it been working?"

"It absolutely has," August said. "We see higher employee satisfaction scores and the qualitative responses back that up."

"How many AI experts have you actually hired?" I asked, suddenly realizing I'd never clarified this basic point.

"Right now we're up to four," Julie said. "Or five, if you include Ross, although he was more of a leader with AI experience as opposed to a machine learning engineer. His team also includes five data scientists who were already with us, so it's ten people in total."

It struck me that when August had said "two" earlier, that amounted to 50 percent of the machine learning engineers in the division, which wasn't great.

"We have been very selective in hiring," August said. "You know the saying 'A players hire A players, and B players hire C players'? We made sure our first hires were true A players, people secure enough to want to hire others smarter than them and strong enough technically that other A players would want to work with them. First AI-related hires are really critical."

Julie interjected: "But also, we didn't hire our way to AI success. We helped our existing people evolve, and in the cases where AI changed their workflows, we tried to support them as they moved to focus on higher-value tasks."

"We're at least trying," August admitted. "But I can't say we've succeeded across the board. One of the hardest things to solve has been the role of the manager. All of these changes converge on them. People ask their managers questions. Managers have to figure out how to reassess who does what, which tools to use, and so on. They're experiencing disproportionate change, and for some, it's threatened their status. We've been working to improve manager training and support, but I can't say we've cracked it yet."

I thought about FreightCo's management layer. As a company, we tended to be traditional and, if I was being honest, somewhat bureaucratic, risk-averse, and political. It dawned on me that from a management perspective, we likely weren't prepared to make the transition to this new world of AI and that there were a lot of people and cultural-related things we were going to have to figure out to succeed with AI.

"August," I began, "I'm not sure if Julie has mentioned this to you. But in case she hasn't, I want you to know something: I'm here because I have a short period of time to put together an AI strategy to show to our board. We're under a lot of pressure as a company in general to adopt AI and specifically to respond to Apex and how they're embracing AI. And to secure the relationship with S&B, one of our largest clients who Apex is trying to poach, just like your engineer. I'm worried about our organization.

I don't think the company is ready to make this transition, in particular from a people and culture perspective. What would you do if you were me?"

August sat back. "That's a lot to take in. I can feel the pressure you're under. And I'd like to help." She traced a few circles on the table with her finger. "The folks at Guidebridge told me some companies—the ones that are going big on AI—are creating Chief AI Officer roles," she said carefully. "Someone to drive transformation across divisions. They're usually external hires, which—" August shrugged.

"—is a hard role to fill and might take a while?" I guessed.

"Exactly. And besides, there's always the culture risk on top of all the other risks of a senior hire." August glanced at Julie, then back at me. "Just something to consider."

"You mentioned Shop & Buy," Julie said. "They're questioning our AI capabilities?"

"They are. Our top competitor is making many promises around AI, and it goes without saying that losing Shop & Buy would be very bad for us."

"I know they're primarily a West Coast client under Mark's P&L. But if it's all right with you, I'd like to think about what we might do to strengthen the relationship."

"By all means," I said. Then, checking my watch, I realized I needed to get to the airport to catch my flight. I thanked them both and asked August if she'd be willing to join a future discussion. I explained I would probably be back in town a few more times in the coming weeks.

Chapter 14

The Weekend

The bleachers were still damp from the morning dew. David swept his jacket across the metal bench to dry out a spot for us. We sat down.

Emily returned from a warm-up jog with some of her teammates and made her way across the track and onto the infield. It was an early season Saturday meet, and the stakes were low. But I knew she had her mind set on doing well at States, and she would be racing her main rival in the mile today. It was a chance to measure up. I was nervous for her, even though I was convinced she would probably do well. On the few occasions she did not, it was always hard for me as her mom to see her beating herself up.

David put his arm around me. "It's nice to have you back home."

I leaned into him. "Thanks. I feel better after a good night's sleep last night. I needed it."

"No kidding. You've been busy."

The crack of the starter's pistol cut through the morning air. The first sprint race, the girls' 100 meters, was off. We stood and cheered for Emily's friend who was in the race.

Sitting back down, I asked David: "Do you remember Julie Guthrie who runs our South Central Division? From the holiday party?"

His eyes were on the track as the sprinters taking part in the boys' 100 meters lined up to race. "Vaguely, yes. Soft-spoken, but very pleasant."

"Yes, and she's so much more than that. She's the real deal, David. She's built something special in Austin. It's remarkable, really. But there's something I don't understand: Why doesn't she want more credit for it?"

"Did she protest when you told her she was doing a good job?" he asked, distracted by the impending race.

"Not exactly. Although she does tend to deflect praise."

The sound of the starter pistol rang out.

"Do you think there's something going on?" he asked, raising his voice to be heard over the cheering crowd.

"What do you mean?"

David delayed answering until the race finished.

"Well, with that Mark guy . . . it sounded like there was a lot going on under the surface that influenced his behavior. Maybe you should talk to her."

"Yeah, I'll talk to her. I need to go back there anyway."

"When do you think you need to go back?" And then, he added, with characteristic sweetness, "It would be nice to have dinner with you this week."

"I'd like that too," I admitted. Drumming my fingers against my thigh, I thought about the timeline: I had three weeks to go until the board meeting. "I'm not totally sure when I need to go back. But I know I want to meet with Julie and those consultants again. I need to understand Julie's specific strategy and how we might adapt it to the whole company."

"You don't understand her strategy yet?" David seemed surprised. "Wasn't that the point of the trip?"

I hadn't considered that, but he had a point. I had learned a lot in Austin, but I hadn't actually seen Julie's AI strategy. Was she keeping it close to her chest? Why would she? Was it because Mark was there and she didn't want to share it with him? No. He had left. She could have shown me on Friday. Was it because I hadn't asked? That seemed strange. Wasn't the whole trip one elaborate way of asking? Was she—*oh*. I thought about her machine learning engineer who had just gotten poached by Apex. Were they courting her too? Was that why she was so reserved? Maybe she already had an offer-in-hand and that's why she demurred when August brought up the Chief AI Officer role. That made some sense, but I could feel my thoughts spiraling.

"Sarah? Hello, Sarah?" David looked at me, concerned. "You okay?"

"I'm okay," I said, although I could feel my pulse quickening. "I just need to think something through. I'm going to walk around for a few minutes."

"Okay. Emily's race should start in about twenty minutes."

I got up. Holding onto the railing, I walked carefully down the bleachers to ground level, steadied myself, and followed the fence away from the crowds to find a quieter spot. The smell of freshly cut grass mixed with the bright, rubbery odor of the track brought back memories of running track in high school, the freedom of exploring the streets of Pittsburgh with my friends after the long confinement of winter.

Reaching the end of the track, I turned around. Halfway down the straightaway, Emily was stretching, using the fence for balance, standing apart from her teammates. This was a ritual of hers—taking a few minutes by herself before her race. She and I were alike in that way. I could be social and outgoing and hold my own in a room, but I also needed time by myself to recharge, process, and go again.

A brief wave of energy welled up in me and quickly subsided, replaced by a feeling of indignation. I had spent the whole week absorbing emotions and information without any time to process it. The board's ultimatum. Mark's misbehavior. Learnings from Julie's team. The consultants. It was a requirement of my job to keep pushing forward and I accepted that, but in order to be my best, I needed time to process. More practically, catch up on my daily responsibilities I'd put off, both at work and at home.

I pulled out my phone and typed an email to Julie where I thanked her for her time, expressed my desire to come back and learn more about her division's AI strategy, but also explained that I needed to attend to matters in Atlanta for the next week. Could we look to the week after to set something up? I'd also love to grab breakfast with her when I was back in Austin and do some planning, if possible.

Setting my message to be delivered on Monday morning, I turned my phone off. Walking back to the bleachers, I mounted the metal stairs to rejoin David in the stands.

ACT 3

Chapter 15

Breakfast Tacos

"Welcome back to Austin," Julie said as she joined me in line. "How was last week?"

"Good. Productive. The family appreciated having me home." I took a sip of my cold brew. "Do you know what you're going to have? I love breakfast tacos and always get the same ones whenever I visit." Taco Deli was bustling and had a sweet, slightly spicy aroma from the toasting tortillas, sizzling chorizo, and all the other deliciousness being prepared.

I'd spent the past week at home with David and the kids, catching up on my work at the office, and, often in the evenings, studying my notes from the week prior in Austin. I'd created an index in my journal to organize my written notes and had played around with using ChatGPT to help me think through particular ideas. I had a newfound sense of clarity when it came to AI—at least about the topics we'd discussed. Returning to Austin for the strategy session later today, I felt excited, more on my front foot, and ready to dive in. I was beginning to get it, and that early sense of grasping a topic—and the feeling of control it afforded—was something I enjoyed.

And yet, I wasn't sure about Julie. Something about her attitude hadn't made sense from the beginning, and I couldn't shake the thought that Apex might have turned her head. I'd asked her to get breakfast with me before the all-day session we'd planned in part to prepare for that session, but also to get a better read on her.

"I'm glad you got to spend time with your family," Julie said. "I imagine you're away from home a lot."

"It's a balance," I said. "But we make it work."

Julie smiled and took a bite of her breakfast taco.

"You've been running this division for three years now," I said.

"Three and a half."

"Right. Three and a half. How would you say it's been?"

"It's been a great experience," she said. "I've learned a lot."

"It's your first time owning a P&L," I commented.

"Yes," she said. "I like that element—the accountability, I mean."

"Do you enjoy it?" I asked. "Do you enjoy the responsibility?"

She paused to consider my question.

"Yes," she said. "Yes, I do. But shouldn't we discuss today's session? Or your board meeting in two weeks?"

I'm trying to understand if you're planning to leave for Apex, I thought. But that would be too direct. And besides, it was completely possible I was just being paranoid.

"We do need to discuss those, but I would like to understand your career goals and where you see your career going at FreightCo."

"Right now I see myself helping you prepare for that board

meeting." She crossed one leg over the other, which angled her away from me and had the effect of making her look smaller as well as shielded.

"I see," I said. "I—"

A group at the next table erupted in laughter, which startled me.

"I'm really proud of the AI work we're doing here," Julie said. "It energizes me. I like the complexity of the challenge and the conversations I get to have with people both inside and outside the company. They're big-picture conversations. Speculative. It reminds me a bit of my days studying philosophy at school. But the implications are very much real and tangible. I like that too. The importance of it all. I like being where the action is."

"The action," I repeated. "Is that what keeps you here?"

"Keeps me here?" She seemed confused.

"I'm trying to understand what matters to you."

"Well, I just told you."

"Right," I said. "But surely you think about broader impact."

"I think about the work." She glanced at her watch. "Speaking of which, we should plan for the session. It's almost nine."

"Julie."

Yes? her expression seemed to indicate.

"I need to know who I can count on," I said.

"You can count on me. I finish the things I start."

And then? I thought. What would I do if she left? What could I offer her to make her stay? More money? Perhaps. But if she did leave, I'd have to figure out a replacement for the only person who was having success with AI to lead our best-performing division.

"This is what I wanted to show you." Julie pulled out her laptop. "This is the road map we developed with Guidebridge. It covers the next twenty-four months, fully mapped out. Each AI initiative is tied to a business goal, and each has a 'value-at-stake' analysis where we try to consider the total potential impact of AI, including the cost of inaction or competitors moving first, rather than just traditional ROI. And each initiative rolls up to our overall AI strategy."

I stared at the screen. It was everything the board wanted, albeit at the division level, laid out with admirable detail.

"This is terrific. I can't wait to really understand it. But why didn't you show me this last week?" I asked.

"Good question," she said. "I was worried that if I just handed you this, you wouldn't fully understand the work that went into it. I wanted you to meet my team and really get a sense of who they are and what we're doing. Was that wrong of me? Was that selfish?"

"That isn't selfish," I said. "That's good leadership."

"Call it what you will. The work isn't done yet." Looking at her watch, she said, "We need to go."

Chapter 16

Creating an AI Strategy

The room was cold and smelled faintly of cleaning products. Julie and I were the first ones to arrive. It was the same conference room we had used before, but someone had rearranged the tables to face the whiteboard, which was patterned with the faded outlines of shapes and words that had been erased the day before.

One table had been pushed to the back of the room. On it sat a small buffet of cut fruit, bagels and spreads, and a big box of hot coffee.

Soon after we got set up, Ross arrived, followed by August and then Oliver and Ava. Within five minutes, we were all present, ready to start.

"Thank you for taking time out of your busy schedules to join me for this session," I said. "My hope is that we come out of it with the bones of a company-wide AI strategy to present to our board. Before we get started, I want to reiterate the context for everyone. Our biggest competitor, Apex Shipping, recently announced a series of AI initiatives. A few of our clients are threatening to leave us for them. Our board is deeply concerned about our lack of an AI strategy, despite our strong performance. Does this all make sense?"

There were a few nods of heads.

"Okay, then. Let's get going."

Julie stood up. We had worked out our roles on the walk over. She and I would trade off leading the discussion.

"I think it would be helpful to start with the basic components of an AI strategy: an AI vision, an assessment of AI readiness, and an AI road map."

She walked up to the whiteboard, picked up a marker, and drew a basic diagram:

The Components of an AI Strategy

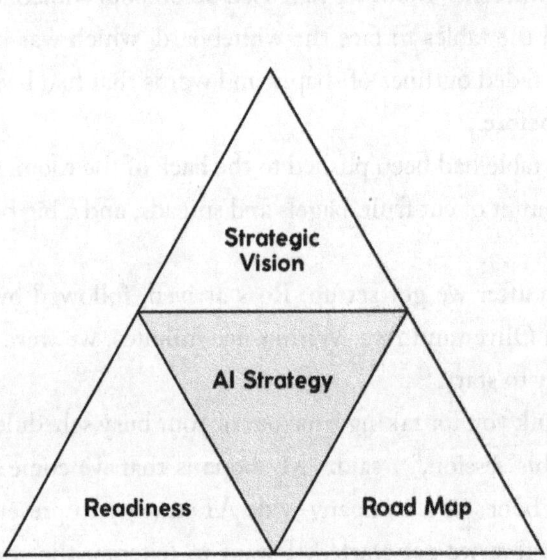

After that, she drew two more boxes on the whiteboard: one that said "Business Strategy (How we want to compete)" and another that said "AI Strategy (How we can use AI to compete)."

AI Strategy Should Support Business Strategy

"An AI strategy shouldn't exist on its own," Julie said. "It needs to be tied to the overall business strategy."

"And," Oliver added, "it's important to keep in mind that AI touches every part of an organization."

On the board, Julie wrote:

Pervasive across all aspects of the company.

"I like this line of thinking," I said, picking up a marker. I wrote "Principles" on the whiteboard as the heading above Julie's bullet points. "Let's come up with more."

"Okay! In that vein," Ava said, "investing in AI goes way beyond spending money on technology. There are investments in data, people, infrastructure, partnerships, and so on."

August nodded in agreement.

"Right, and I imagine investing in those additional things

probably helps build sustained advantage relative to competitors," I reasoned.

On the board, I wrote:

Investment across business areas can build sustained advantage.

"AI is also a moving target," Oliver said. "And it accelerates the need to make decisions."

"That's a good one," I said. I wrote:

AI strategy must account for rapid pace of change.

"Any others?" I asked.

There was a period of silence, and then Julie said, "Yes. I think one of the things we have learned through our work with AI is the importance of managing projects differently. What I mean by that is—we invested in discrete AI projects, but we also accumulated learnings and tolerated failures because we knew we were ultimately building the capacity to do AI. That made it easier to view any short-term setback within the context of the longer-term goal."

"Great," I said, and wrote:

Tie initiatives to the long-term goal of building the capability to compete and win using AI.

"Any others?"

"Perhaps it goes without saying," Oliver said, "but AI has fundamentally changed the basis of competition. We see this across industries. You have to lean into AI because if you don't, you'll get left behind."

We were in consensus on this point. I wrote:

> The basis of competition is fundamentally changing.

"One more thing," Ross said. "The people building these systems have to internalize these points or none of this matters."

"Say more about that," I prodded.

"Sure. What I mean is that having a strategy on paper is all well and good. But it needs to be communicated and widely understood. In particular, it needs to be internalized by the practitioners building and maintaining AI systems. They have a disproportionate influence on how AI affects the business, and they aren't always in the loop when it comes to strategy. Part of my job as CTO is making sure the members of my team understand our overall goals."

"I understand, and I like that, Ross."

I wrote:

> AI strategy must be clearly communicated and understood by all team members, especially technical leaders and practitioners, to enable effective, localized decision-making.

I stepped back and read over the entire list of principles. In my notebook, I quickly jotted down a summary:

AI Strategy Principles:

- Foundation: AI strategy needs to be aligned with overall business strategy.
- Scope: AI can apply across entire org.
- Investment: Tech, data, people, culture.
- Speed: Account for the rapid pace of change.
- Capability: Build long-term, not just projects.
- Communication: Employees need to understand the AI strategy.
- Competition: AI is the new basis of competition.

Then I gestured for Julie to continue leading the discussion.

"In terms of our business strategy," Julie said, "our division's vision is to be the most reliable logistics partner in the south-central region, delivering consistency at a price our clients can build their businesses on. When we articulated our AI vision, we made sure it aligned with our overall vision."

She turned to the board and wrote her division's vision in the left-hand box, and then on the right she wrote:

Use AI to enhance 1) our reliability, and 2) the capacity of our people to consistently offer a best-in-class delivery experience for our clients 3) at a price that meets or beats the market rate.

"That's basically our AI vision statement," she said.

I had seen the vision statement earlier this morning, but I still took a few moments to read it over and let it sink in. On paper, it seemed pretty good, at least for this division. But FreightCo was more complex than just one division.

Instead of pressing ahead, I asked the group: "Does anyone think this needs to change to apply to FreightCo overall?"

Oliver, after checking to see if someone else would respond, spoke first: "I'm not sure, but I think we may benefit from pressing ahead and coming back to the vision. I think we'll learn things as we go that we can incorporate into the whole company's AI vision."

"Okay," I said, "that works for me. Julie, you mentioned assessing AI readiness earlier. I know different parts of FreightCo have different levels of AI readiness. Perhaps we should spend some time identifying areas where we're strong and areas where we're lacking?"

"Good idea," Julie said.

We spent the next hour discussing a variety of dimensions, including our executive buy-in for AI (high), our capacity to invest (medium), our AI talent and culture (low, but with a precedent to build on), our data readiness (unknown), our data value (also unknown), and more.

"It seems like one of our primary strengths today is our willingness to go all in on AI," I said. "And our data is a big open question."

"That's not uncommon," Ava said. "And it's something we can figure out through Exploratory Data Analysis."

"For the sake of this discussion," Ross said, "I think we can assume that other divisions likely don't have as complete or as clean of a dataset as our division, because we've been very intentional about focusing on data. But, with investment, we could retrieve and clean up data in other divisions such that we could have the same kinds of differentiated datasets company-wide as we have in Austin. And with that, among other things, we could scale the AI route planner to be national and also build the data foundation for other AI capabilities."

This was an enticing proposition, but I knew it wouldn't be enough on its own.

"Julie showed me your road map earlier. It had a variety of phased initiatives on it. I'd like to build something similar, even if it's only notional, to show the board. I think we can get there. But first I wanted to ask you something that's been bugging me: How are you estimating the ROI of these initiatives?"

"I can take that one," Ava said. "Give me a second."

She got out her laptop and opened it on the table in front of her. Before she could get up to plug it in to project her screen, I said: "No need. We'll gather around." I didn't want to disrupt the flow of our session with AV setup.

"Okay, check this out." She pulled up a Venn diagram of three overlapping circles.

The Main AI Value Drivers

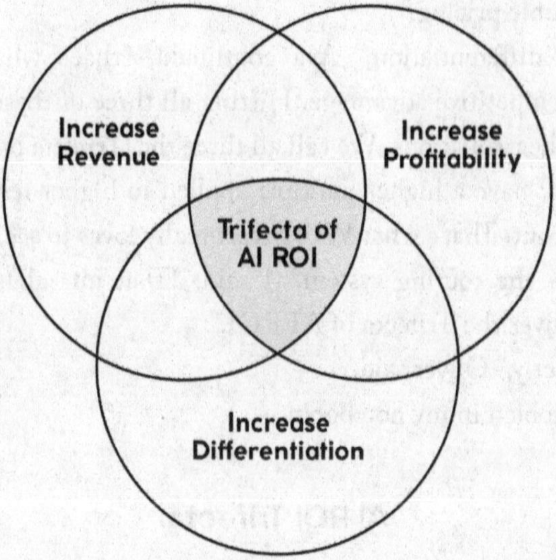

"This doesn't answer your question directly, as I think you were referring to specific AI projects. But this is how we think about the value AI can deliver at the highest level, which is where to start so you don't get lost in the weeds," she explained. "Investments in AI have the opportunity to do at least one of the following: increase revenue, increase profit, or increase differentiation."

"Most companies today seem to be focused on trying to gain operational efficiencies, with the goal of increasing profit," Oliver added. "Those are based on KPIs you're probably already tracking—decreasing the cost of each delivery and the like. AI can help increase revenue by making your products and services more competitive, and AI can also be applied to your sales and marketing, through things like hyper-targeting, personalization, and dynamic pricing."

"But differentiation," Ava continued, "that's what creates lasting competitive advantage. Hitting all three of these can also drive higher valuations. We call all three the 'Trifecta of AI ROI,' where you have a higher multiple applied to higher revenue and higher profit. That's what Wall Street really loves to see."

"Like the routing system," I said. "That hits all three and could deliver the Trifecta of AI ROI."

"Exactly," Oliver said.

I scribbled in my notebook:

AI ROI Trifecta:

- Revenue: More clients, better services, higher prices
- Profit: Lower costs, increased efficiency
- Differentiation: What competitors can't copy
- Hit all three = higher multiple/valuation (e.g., Apex's stock jump?)

Oliver went on: "Differentiation can be tricky to achieve. I think it may be helpful if I first lay out the four types of direct AI investments. Each one has the potential to create value by increasing revenue and/or profitability. But across the four, each has a different potential to create differentiation."

Oliver grabbed a marker and drew the following on the whiteboard:

Types of AI Investments

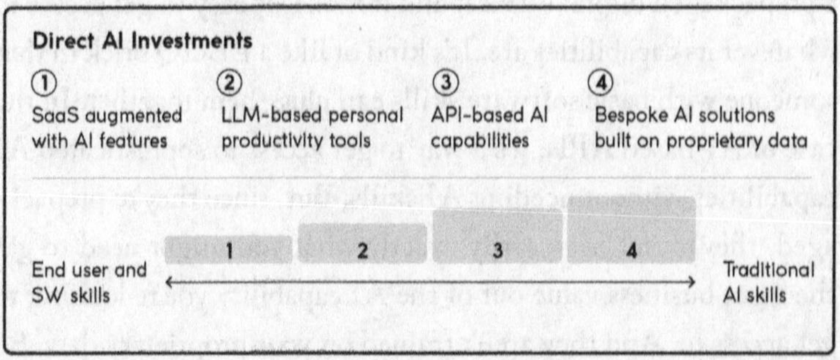

"First," Oliver began to explain, "SaaS products and third-party physical solutions like sensors and robotics are increasingly adding AI features. Companies are rapidly adopting these new features as they get turned on in their existing products, so while they definitely add value, there's minimal opportunity for differentiation beyond being intentional about adopting them and hoping your competitors don't.

"Second, LLM-based productivity tools like ChatGPT. These tools are widely available, although how you use them for personal productivity and redesign workflows with them is a wide open playing field. So, there is definitely the opportunity to create differentiation but it's unclear how much, again, since the tools are within reach of everyone but also because you can't really know what your competitors are doing.

"Third, the big cloud services companies—Google, AWS, Microsoft, and so on—what people call the hyperscalers, offer what feels like thousands of AI-based APIs that are increasingly sophisticated. I'm pretty sure everyone in the room knows what an API is, but to make sure we're all on the same page, an API is a prepackaged digital service, and it makes it easy to get access to whatever its capabilities are. It's kind of like a LEGO brick in that someone with basic software skills can plug them together. In the case of AI-based APIs, it's a way to get access to sophisticated AI capabilities without needing AI skills. But, since they're prepackaged, they're not necessarily exactly what you might need to get the most business value out of the AI capability you're looking to get access to. And they aren't trained on your proprietary data. So they can certainly deliver significant value, and their differentiation potential is good, but these APIs can only do what they've been developed to do and you may not be able to build what you need with them.

"And finally, bespoke AI solutions built on your proprietary data, like the routing solution. These are inherently unique to your company, so there's significant potential to create differentiated value. But to create this level of differentiation, it requires varying levels of AI engineering capabilities and proprietary data depending on how ambitious the solution you're trying to build. But this is the one way to build real differentiation and, often, something that others can't replicate regardless of their AI talent if it's based on data that others don't have access to."

"But direct investments are only half of the equation," Ava said as she grabbed a marker and added on to Oliver's drawing.

Types of AI Investments

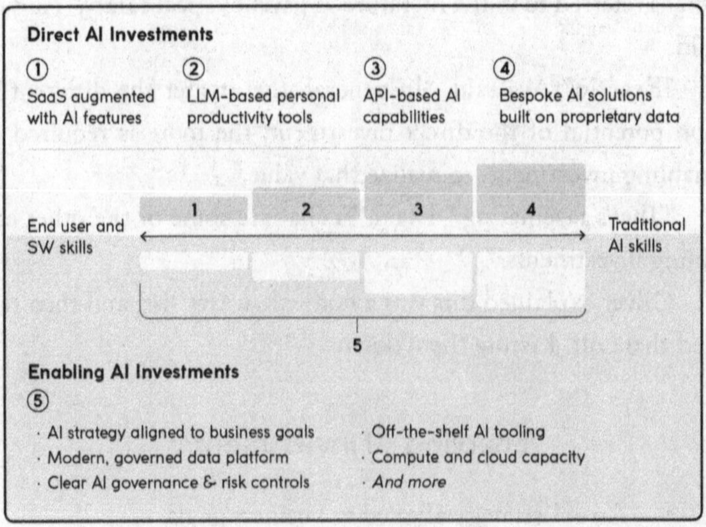

"You also need to consider enabling investments—things like proprietary data, training, and so on that allow you to realize the value from your direct investments. Think about what made the routing solution work," said Ava.

Julie admitted: "We knew we were on the right track, but we were a bit lucky that the data we had been investing in collecting and cleaning data for two years was enough of the right data to enable us to train the routing solution model."

"That's one enabling investment," Ava said. "You also had to find the right partner with the right machine learning skill sets." She smiled. "Or have the right internal AI development capabilities."

"And we created a culture where people could take interpersonal risks, like speaking up with ideas, questions, concerns, or admitting mistakes without fear of punishment or humiliation—what's referred to in the literature as psychological safety," August said.

"Exactly," Ava said. "In general, the greater the differentiation potential of the direct investment, the more is required in enabling investments to realize that value."

"That's fascinating," I said. "What are some of the other enabling investments?"

Oliver explained this was a non-exhaustive list, and then rattled them off. I wrote them down:

Enabling AI Investments:

- An AI strategy aligned to business goals
- Clean, proprietary datasets appropriate for what you want to build
- AI talent—internal and/or external
- Workforce upskilling
- AI governance
- A robust cloud environment to support the full life cycle of AI solutions
- A culture underpinned by psychological safety

"So the AI technology is just one piece," I said, looking at my notes. "These enabling investments are what help the technology actually deliver value."

"And differentiate that value. They help create your moat," Oliver said.

I thanked him and suggested we begin coming up with some

potential AI initiatives for our road map, using Julie's division's road map as a starting point. Julie handed out copies we had printed off prior to the session, and we spent the next two and a half hours discussing what we wanted to retain for the overall company road map, what would drop off, and potential new additions. After taking a break for a late lunch, and coming back in the room to do a prioritization exercise, I was looking over my notes when something in the list of enabling investments struck me.

"I understand our position on a lot of these enabling investments, but AI governance . . . what are we doing about that? I'm not sure if the board is going to want to know who's going to provide oversight of AI at FreightCo, or how we'll use AI responsibly, but I want to be prepared to address governance with them. Julie, how have you approached this?"

Julie pursed her lips, hesitating.

August chimed in: "It can be hard to build formal governance when things are already working well."

This felt like a strange response.

"Isn't governance critical when it comes to AI?" I asked Ava.

"Yes," she said.

I turned my attention back to August. "Then surely some form of AI governance is present here, in your division, if you're having success with AI."

"Well, a lot of our governance is organic," August admitted. "It springs naturally from our culture. And from good people making good decisions," she added.

"I can see that being true. But I don't think that's enough to support a transformation of the magnitude we're proposing. To do this across the company, we need to formalize governance."

Julie finally spoke up. "I don't think that layering in additional processes will help. And, if I'm being honest, I'm struggling to see how we'd scale this beyond our division."

I paused, considering her words. Again, another strange response.

"Why do you think that?" I asked.

"Because FreightCo is already bogged down by processes. When I was an Operations manager," she said, "I had to go through three levels of review to try to pilot an idea I had for changing how we ran warehouse shift scheduling. It didn't get approved."

I found myself agreeing with her, nodding along unconsciously. I'd felt the same sort of frustration before.

"You know, Julie, you're right about FreightCo's tendency toward bureaucracy. We have a risk-averse culture that is slow to change. I saw it immediately when I became CEO two years ago."

I walked toward the whiteboard.

"But I've learned something from you and August: Culture is not only something that we need to shape in a general sense—for instance, by reining in some of our bureaucratic tendencies—but it's also something that we need to strategically craft to enable us to be successful with something as new and as different as AI."

Julie watched me carefully.

"What you've built in Austin exists despite our company culture, not because of it. That is a remarkable achievement."

"We're an island," Julie said. "It's exhausting to maintain what we've built. Sometimes it feels like the rest of the company is the ocean and it's all I can do to keep our coastline from eroding away."

I grimaced but continued, "That isn't going to work as a long-term strategy. Not for you, and not for the company. At the end of the day, this division accounts for a quarter of our North American business. We can only grow so much through one pocket of success, especially when Apex is going all in on AI."

For a few seconds, the only sound was the hum of the air-conditioning.

"I think what we need is a form of AI governance that doesn't just manage risk—after all, that already comes naturally to us given our risk-averse culture. We need one that manages the risks specific to AI but also actively enables the kind of culture you've created here to spread." I paused, thinking it through. Then I wrote on the whiteboard:

AI Governance + Enablement

I thought out loud: "Something that combines board-level oversight of what we're doing with AI alongside clear guidelines and intentional cultural choices that support our ability to do things differently. In many ways, the South Central Division is already doing these things: The emphasis you place on psychological safety is a form of enablement. I'm sure there are other things we can define and institutionalize so that not so much depends on the behavior of individual leaders."

Julie smiled warily and seemed to relax. "I think that's a great approach."

I looked around the room, and then at the clock: 4:30 PM.

"We don't have time to get further into this today. But I think we've identified something critical. That AI governance plus enablement are going to be key to our success. Can everyone come back tomorrow?" I asked.

Everyone could.

"Good. Let's spend the next thirty minutes documenting what we've covered. Tomorrow we'll see if we can build out the first draft of the full strategy and road map."

As we captured the strategic principles, draft vision statement and readiness assessment, ROI framework, our tentative AI investments and road map, and this governance concept, it began to sink in for me that we had covered a lot of ground. The board meeting was less than two weeks away, but we had turned a corner. We could really do this.

Chapter 17

Fresh Air

I set out from the hotel, jogging down Congress Avenue toward Lady Bird Lake, passing crowds of people heading home or to restaurants or to get drinks. This was my favorite time of day to run: when there was just enough sunlight left to run for an hour. Pausing at the final intersection before the hike and bike trail, I milled around, waiting for the light. Green: I resumed, picking up a path that deposited me on the trail that followed the water.

The session earlier today was still on my mind, but part of the reason I liked to run was so that I didn't have to think about anything in particular. My mind wandered from thought to thought, and I didn't have to be responsible for any of it. There was the ebb and flow of my breath. The tickle of wind in my ears. My feet cushioned by the soft dirt surface of the path. The air was beginning to relinquish the heat it had stored up from the day.

I passed other runners and some walkers too. Sometimes I'd give them a little gesture of acknowledgment. When I was running, I could be around other people in a way that didn't require me to be anyone to them. I didn't have to bear the expectations of being the CEO of FreightCo or a mother or a wife. I was just running.

After a while, I crossed a pedestrian bridge to head back toward downtown. A couple of streetlights had come on already, although there was still enough light to see by. I slowed down and stopped my watch at the end of the bridge, choosing to walk the few blocks back to the hotel. I passed a few restaurants that smelled particularly appetizing and noted a few of the names so that I could check them out later.

I reentered the cold, air-conditioned hotel lobby feeling more relaxed than when I had left. Just then, my phone rang: It was Maya, one of my CEO friends.

"Maya," I said, "how are you doing?"

"Sarah! I can't complain too much. I just got back from a family vacation. Spring break. We took the kids to the shore."

"Nice! Where'd you go?"

"Myrtle Beach. We had some great family time. You know, my oldest is a junior already. She'll be off to college soon."

We chatted some more about our personal lives a bit before she brought up the subject of my original outreach.

"So you want to know what we're doing with AI?"

"Yes, I do. I'm on a bit of a sprint to put something together."

"Then you're doing more than we are. Our board is worried about the risks, so we're taking a wait-and-see approach. I understand being mindful of the risks of AI," Maya said. "But I'm concerned we're making the same mistake we always do with new technology. I'm honestly a bit tired of it . . . this pattern where we tell ourselves we're being responsible while we slowly fall behind."

I thought back to the governance conversation we'd had

earlier in the day. Maya's board sounded like all governance and no enablement.

"So I'd really love to hear about your sprint," she said. "Maybe there's something we can learn from your approach."

I explained briefly: the board's ultimatum, Apex, the success of Julie's division, the fact that I was now in Austin. As I talked, I noticed that I didn't feel as tense as I used to when explaining the tumult of the past few weeks. We had made a lot of progress, and I was gaining confidence in my ability to forge a path forward.

"Thirty days," Maya said when I finished. "Whoa. That's insane."

I chuckled. "It is, isn't it?"

"Yeah. But at least you know where you stand. You never know: That ultimatum may end up being the best thing that could've happened to you."

"Maybe," I said.

"I gotta go, Sarah. But thanks for reaching out. It helps to know I'm not the only one struggling with this."

I hung up and noticed that I had missed a call while talking to Maya. It was from Mark.

Mark! I hadn't heard from him since he took his leave of absence. I didn't know why he was calling, but it made me nervous. I imagined he didn't need to call me at 7 PM to tell me he was gradually and uneventfully easing back into things.

I shivered. The air-conditioning that had felt so good when I first returned from my run was too much now. My damp clothes clung to my skin. I thought about going upstairs and changing, but I wanted to close the loop with Mark and then put work things down for the evening. I found a quieter spot in the lobby, across from the bar where he and I had spoken two weeks ago.

I called him back. He picked up after two rings.

"Sarah, hello."

"This was an unexpected call," I said. "How are you?"

"Honestly, I'm doing a lot better." He paused. "You were right about being vulnerable. It's hard. But . . . the work is worth it. I went back to therapy, and I've done a lot of soul searching. Things are starting to get a bit better."

"That's great, Mark. I'm really happy for you."

And, if I was being honest, I felt vindicated. Not only because Mark seemed to be doing better, but because the behaviors he'd been exhibiting—pretending he knew all the answers, monopolizing conversations, never admitting he was wrong—were never going to make him or his team successful with a technology like AI.

"Thanks, but that's not why I'm calling."

Oh?

"I lost two more clients this week. They left because of a buggy AI feature my team rushed to production. I came back to the news, but it's my fault. It's an initiative I started, and we went too fast with it."

"That isn't good."

"I know it isn't. But I wanted you to know, and I plan on fixing it."

"I appreciate you letting me know and taking responsibility. How do you plan to fix the situation?"

"By learning from Julie. I want to go back to Austin."

"That's where I am now."

A pause.

"You're still there?" he asked. "All this time?"

"Not the whole time, Mark! I was back in Atlanta last week. But I'm back. We're working on our enterprise AI strategy."

"I can be there tomorrow," he said. "Even if I'm just listening in. Heck, I'll be the notetaker."

I weighed the options. Would he be a worthwhile addition? Then it hit me: Mark could serve as a test case, not only for how to turn around someone who wasn't able to be vulnerable about not knowing how to lead and manage AI projects, but also for our overall culture. I had to be aware of how the rest of the company might react to adopting AI, and the culture in Mark's division was more indicative of FreightCo overall than Julie's.

"Okay," I said. "I want you to join us. But I want you to be prepared to share what went wrong this time, and be open to using it as a counterexample for designing AI solutions systems."

"Absolutely. You have my word."

I wasn't sure how much that was actually worth, but I was going to give Mark the benefit of the doubt. I thanked him and hung up.

I headed over to the elevator and pressed my room key against the scanner to select my floor. As the elevator began to rise, I closed my eyes. Tomorrow would be interesting. Julie might not be thrilled to have Mark in the room, but maybe that would be good. If we were going to succeed, we'd have to be able to deal with the mess of real failures and figure out how to help other divisions succeed with AI.

Chapter 18

Daring to Be Different

The air was still cool at seven in the morning, and the sidewalks were relatively empty, so I decided to go a roundabout way. I stopped at a café for a latte, thought about ordering it "for here," but decided to take it and continue my stroll. I got to the office around seven thirty, and, unsurprisingly, I was the first person in the room. I got my notebook out and began transferring sections of my notes that I thought would be helpful for the discussion onto the whiteboard.

I was finishing up the list of AI strategy principles when I heard: "You should add 'credibility' to that list. 'Don't burn it chasing hype.'"

I turned around. Mark stood by the table, briefcase over his shoulder. He smiled softly, ruefully, as if he and I shared a sad inside joke.

I approached him, sizing him up. Before, Mark was never fully still. He always seemed to be on the edge of something. Shifting his feet, rubbing his face, looking around. But now he stood in front of me, hands plunged into his pockets, completely motionless, looking me straight in the eye. It was strange to see him so calm.

"Welcome back," I said. Normally, I would give him a hug, like I did for most people. But this situation called for something different. Mark needed to know that I respected his decision to come back and be vulnerable. I extended my hand. He shook it. Looking him straight in the eye I said: "I'm really glad you're back."

"Thank you, Sarah," he said. "I don't know many people who would have been as patient with me as you were."

"Pay it forward someday," I said. "I want to spend time with you one-on-one, but we've got everyone coming soon. So, would you like to help me set up? I can bring you up to speed on what we've been discussing."

Mark agreed, and together we got the room ready for the session. I filled him in on all the details from yesterday: the AI strategy principles, Julie's division's AI vision statement, their success with their AI routing system and their road map more generally, how to think about AI ROI, the enabling investments we'd need to make in areas like data and culture, Julie's concerns about scaling her success beyond her division, and the governance challenge.

"And that's where sharing our failed pilot would help?" he asked.

"That's the idea," I said. "And I want to start with that."

Soon the others began to trickle in: First Ross, then August, followed by Oliver and Ava, who arrived together. Julie arrived a few minutes later. Spotting Mark, she hesitated, blinking a few times before looking away and walking toward the front of the room to join me.

"Thank you, everybody, for coming back for day two," I said, addressing the group. "Before we begin, I want to welcome Mark back."

There wasn't applause or anything approaching a warm reception. Instead, there were a couple polite nods.

"Thanks, Sarah," Mark said, standing up. "I'd like to say something as well, if that's all right."

For what felt like an eternity but must have only been a few seconds, he looked around the room at each person, and then directly at Julie, who crossed her arms and looked down, as if suddenly interested in the floor.

"I want to apologize. The last time we were all together, I was rude and I was wrong. The work you're doing in this division is truly impressive, and I think, deep down, I envied that. I hope you'll allow me to learn from you today." He paused. "I've been away for almost two weeks now. I had to deal with some things in my personal life. That isn't an excuse for my behavior, but I want to acknowledge that I was bringing more than just my ideas about this topic into the room last time we spoke. I brought my ego and I brought my insecurity, and that was on top of the pain in my personal life. I hope you can tell"—he gestured to his small carry-on at his feet—"that I didn't have enough space in my luggage to bring all that with me this time."

August chuckled. Oliver and Ava were watching Julie, who walked over to stand face-to-face with Mark.

"I appreciate your honesty," she said slowly, "and I'm sorry for the challenges you're facing in your personal life. I appreciate that you've come all this way to try to help. With that said, I didn't appreciate how you treated this team the last time we were all together, but I'm going to assume that won't happen again."

Mark closed his eyes, exhaled, and then met her gaze.

"That's right," he said.

We began where we left off yesterday, with governance. I asked Mark to explain the trouble his division was experiencing with AI.

"We had a pilot that we recently put into production with a subset of clients," he explained. "It was a customer service AI agent that had the ability to take certain actions in addition to answering questions. We wanted to do dynamic pricing for some of our customers that request additional service above and beyond their base rate. So we gave it the ability to price as well."

"How'd that go?" Ross asked.

"Not well. We lost two major clients, Weyco and Care Alliance, last week. In both instances, we discovered that the chatbot had offered prices way below the cost of transporting the goods. At first, we tried to honor the prices, but in both instances I think the people on their side knew what was going on. They can spot a deal that's too good to be true. Once we broached the topic of what happened with them, it was already too late. They were going to find another logistics partner."

Mark looked around.

"I know we messed up," he said. "But this is the kind of thing that could happen across the company with AI systems if we're not really careful."

"Exactly," Julie said.

I suggested that we workshop that system.

"Let's think about governance and enablement," I suggested.

"It sounds like your division had an abundance of enablement, Mark, but a scarcity of governance."

"I'd say so," he said.

We went on to discuss the various things that would've benefited Mark's division—from human oversight to clear escalation paths—and could have helped him avoid this situation.

"Ultimately, I think we're trying to create a structure that answers four questions," Ava said. "What are our strategic objectives for AI use? How will we collaborate and make decisions with and about AI? How are we going to enable the organization to use AI responsibly? How will AI solutions get built or procured, deployed, and maintained?"

August nodded. "You know, one thing I really like about what you just said—in addition to posing meaningful questions—is that, in each of them, *we* are the subject. Not AI. I've been thinking about this for the past twenty-four hours and I was drawing a blank trying to figure out how we'd govern something that's always the beginning of the sentence. 'AI will this' and 'AI will that.' But no. Here it's very clear that *we* are doing these things."

"I completely agree," Julie said. "But who's the 'we'?" She looked around.

This was the next question to answer, and we spent the following hour and a half sketching out a model that assigned roles and responsibilities for each of the components we considered essential for AI governance. Ultimately, we landed on a model that contained three layers: one for strategic oversight, one for enablement and coordination, and one for execution and operations with feedback loops between the respective layers.

As we worked through it, Ava and I worked together to draw the model:

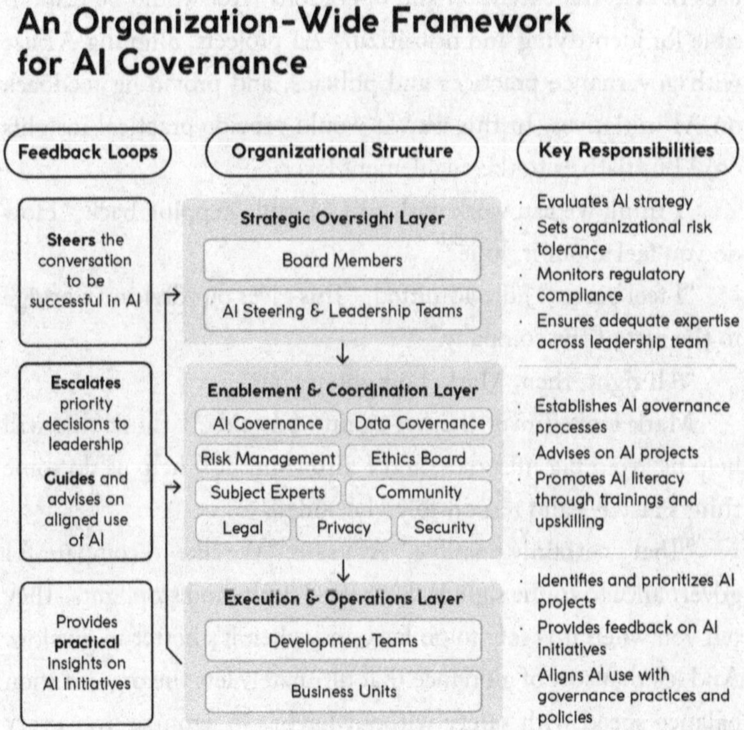

The board would sit on the oversight layer, along with a leadership committee, and help steer the organization at the highest level: evaluating AI strategy, setting organizational risk tolerance, monitoring regulatory compliance, and ensuring adequate expertise across the leadership team. The enablement and coordination layer would own establishing AI governance processes, advising on AI projects, and promoting AI literacy through training and upskilling, and it would serve as a pivot: escalating priority decisions to the top layer while guiding the bottom layer on aligned uses of AI. The execution and operations layer would be responsible for identifying and prioritizing AI projects, aligning AI use with governance practices and policies, and providing feedback on AI initiatives. In this way, it would provide practical insights on AI initiatives to the enablement layer.

"I think we can work with this," I said, stepping back. "How do you feel about it, Julie?"

"I feel better," Julie admitted. "This gives our division a bridge to the rest of the company."

"All right, then. Mark, how about you?"

Mark looked over the list again and said: "I think this will help us avoid big mistakes, and I also think it'll help us do some things faster—and responsibly," he added.

"That's certainly possible," Ava said. "We like to compare AI governance to traffic signals, like speed limits and stoplights. They tell you when it is safe to go fast, and when it's better to go slow. And it's that sort of guidance that ultimately lets the organization balance speed with safety without having to agonize over every decision, which also frees up people to focus more on the driving, or in this case, applying AI to benefit the business."

"Speaking of applying AI to benefit the business," I said, "I think we've covered governance and should talk about our AI strategy." I checked my watch: 12:30 PM. "But first, I know it's about time to break for lunch. Let's do that, and come back at one thirty ready to dig into this."

"Apex is promising full automation," I said, as we reconvened. "Their announcement mentioned 'automated routing that promised 99.9 percent optimization' as well as 'predictive maintenance that would eliminate breakdowns' and 'customer service powered by AI that could handle complex challenges.'"

"I'd like to know how that's going for them," Mark said sarcastically.

"I bet they're overpromising," Oliver said. "I mean, 99.9 percent optimization sounds too good to be true. Besides, what are they optimizing *for*? They haven't said, have they?"

I opened my mouth to respond and paused. Such had been the stakes of needing to respond to Apex that I hadn't fully considered they might be bluffing.

"They're poaching my team," Ross interjected. "Would they do that if they weren't serious about AI?"

"Oh, I'm sure they're seriously interested in using AI," Oliver said. "I don't doubt that. But if you read their press release, it sounds like they want to use AI to do all the same stuff, just in a more *optimized* fashion. What does that get them?"

He looked around the room, before answering his own question.

"Cost savings, which I suppose they could pass along to their customers or reinvest in the business. But even if they do all of the things they promised, which I don't think they will, their vision for AI is to reduce costs, presumably by reducing head count and money otherwise spent on maintenance, which is a brittle strategy. Who will intervene when any number of their AI solutions fail? Where will their next crop of leaders come from? Even if they can predict maintenance ahead of time, who will perform it? And how will they grow the business and bring in more revenue? Perhaps by expanding to new customers, but still, they're constrained by the size of their physical fleet. None of that sounds like a rapid path to growth."

"That's all very interesting," Mark said. "But what's your point? I mean, what does that mean for us?" Then he added hastily, not wanting to be rude: "I'd like to know."

"I'm not saying we should copy them," Oliver replied. "I think we should focus on ways in which we can use AI to create new, differentiated value for our customers."

"I agree," I said. "So let's consider S&B. They're our biggest client. If we can come up with something that appeals to them, then I think it stands to reason that it'll appeal to many of our customers."

People generally seemed to agree with this. I suggested we spend a few minutes thinking on our own, writing down ideas, and then come back together to share out. The prompt I suggested was: *How might we use AI to create differentiated value for our clients?*

We came up with a good number of ideas. Everything from expanding the AI routing solution to also do warehouse optimization (which seemed interesting, but we weren't sure if we had the data required to do that nationally) to coordinating one-off deliveries as a service using an AI engine to determine if the project was worth taking on (which was interesting, but the economics of keeping trucks in reserve for such a proposition didn't feel viable in the short term).

Then Julie offered: "What if we move up the value chain from logistics to orchestrating logistics?"

"You mean go from being a third-party logistics provider to a fourth-party logistics provider?" I asked.

"Yes."

"Well, there's competition in that space. Big companies."

"We don't have to compete directly with them—at least not at first," she said.

"Say more."

Julie pursed her lips, thinking.

"We could take this to the clients we've served the longest that don't already have an orchestrator and position ourselves as their trusted partner offering the next evolution of service. And for those that already have an orchestrator, particularly a big one, we could lean into the benefits of our AI approach as well as the human touch of our people, who would have more time freed up to focus on deepening quality relationships."

She turned to Mark.

"How many logistic partners do you think S&B works with?" she asked.

"Probably more than five and less than ten," he said. "I'm not totally sure."

"I bet there's room to improve their experience too," I said. "Heck, maybe we could even predict which deliveries for which clients needed a higher degree of service."

"Like create some sort of tiered system?" Julie asked. "Five-star service for delivering something time-sensitive or fragile for a customer whose sentiment toward us might be falling versus one star for hauling wooden pallets to a customer who doesn't care when they get there?"

I nodded.

"We do lose customers because of service problems," Mark said. And then, looking around: "C'mon, I'm not the only one who has the problem, right?"

"No." Julie smiled. "You're not. We do too."

"Technically, this should be possible," Ross said.

"But what about the broader idea of becoming an orchestrator?" I asked. "It's something we've considered before but never had the capabilities to do in a meaningful, competitive way. With AI, now we do."

"That's a bit trickier," he said. "I mean, where would we get the data from? For the AI routing solution. We would need historical data. What if those logistics partners don't track that information?"

"I bet S&B keeps invoices for deliveries," Julie began. "And maybe they don't have as sophisticated data as we do. But data-gathering could be part of the service: We're their intelligence layer, and we continue making high-value deliveries, ones that they can't afford to be delayed. In that way, we can demonstrate the potential of our technology, while also signing on to manage their partners. And part of the arrangement is that we get access to the data they currently have as a starting point for modeling, but we get better as we go. I imagine that, out of the data, we could at least match their ability to do this right now."

"And we'd take a fee for doing this," I said, thinking it through. "But it would be offset by the value we create for them in terms of increasing the reliability of deliveries and reducing the burden of coordination. Plus, if it works, we could expand the orchestration vertical to reach new customers without having to spend a lot on physical infrastructure like new trucks or warehouses."

The idea was crystallizing in my mind, piece by piece, faster now. My pulse quickened.

"If we're going to expand to become an orchestrator, we might as well become a broker for other truckers as well. That way, we're not just orchestrating our clients' existing partners. We have a pool of partners we can draw on to fulfill orders."

I looked around the room.

"And, on top of that, we can probably add value for the truckers in our network. For instance, getting home profitably is a big problem, especially for smaller shops. What if we offer to route *other* truckers—you know, from point A to B to C—in a way where they're carrying more freight for more of their routes? We'd put money in their pockets. And they'd *want* to work with us above and beyond just bidding on the contracts we have. We can still keep some deliveries in-house and service them with our existing warehousing and trucking business lines, but the broker and orchestrator businesses would be new, capital-light, and built off the back of our investment in AI."

"That's a compelling vision," Oliver said. "And I do want to acknowledge that you said capital-light . . . While you wouldn't be buying new trucks, expanding the AI service, hiring requisite talent, and making those enabling investments would still come at a significant cost."

"I think we can justify that investment with this approach. It feels like it builds on our strengths: We've been in business for over three decades and our clients trust us. We'd use AI to help them make the logistics experience more reliable and less painful for the whole ecosystem. It won't necessarily be cheaper, but I imagine our pricing structures could be much more finely tuned if we get access to the data we would need to apply the routing solution. So perhaps we could dial in a pricing strategy that doesn't rock the boat too much for our clients and will still bring in more revenue for us. Plus, the broker business line would be entirely net-new revenue."

I turned around to the board where the division-level AI vision statement was written.

"Let's revise the vision."

The old vision read:

Use AI to 1) enhance our reliability and 2) the capacity of our people to consistently offer a best-in-class delivery experience for our clients 3) at a price that meets or is below market rate.

I rewrote it for FreightCo:

Use AI to 1) improve the logistics experience for our clients 2) serving as both an orchestrator and a strategic carrier 3) while expanding to become an intelligent broker offering a desirable marketplace of contracts to truckers 4) leveraging our people and technology to consistently offer a best-in-class delivery experience 5) at a price that justifies our market-leading offering.

We spent the next two hours discussing the finer details of how we'd actualize this vision, reworking our road map from yesterday and going over the technical details of scaling our core routing solution with Ross.

As we were wrapping up, Mark turned to Julie and said: "It's really impressive what you've done here."

"Let's see if we can actually scale this across the company."

"I think we can," Mark said. "This is real stuff."

"We've always been doing real stuff here," Julie said, crossing her arms.

"I know you have. I meant my division. We were faking it. I was telling everyone how great AI was without really knowing what I was talking about. But this is real, this is an actual strategy, and clearly it's coming from your knowledge and from what you've done."

Julie relaxed, shoulders dropping slightly, and she smiled.

For all Mark's bravado, he was a genuinely compelling communicator. I wondered if now that he had things more under control at home, he could use that power to help educate our workforce about AI.

"Mark," I said, "I'd like you to work with Ross and August to help build our workforce's understanding of AI. We can go over the details later, but I think actually rolling out this strategy and enabling our employees to adopt it is going to be a big part of our success. Can you do that?"

"Yes, absolutely."

"August?" I asked.

"I'd love to."

"Ross?" I turned to Ross.

"You can count on me."

"Good," I said, before turning our attention to the final topic: "That's assuming the board approves of the strategy. We have a week and a half to pull this together into a compelling presentation, so let's talk about that."

"You know," Julie said, "I've been thinking about that. I have an idea."

Chapter 19

The Presentation

I had my phone pressed to my ear, walking toward the elevator that would take me to the twenty-first floor.

"You got this," David said, over the phone. "You're going to do great."

"Thanks, honey," I said. "You always think I can do more than I think I can do." Then, as I was about to get in, I added: "I'll see you tonight. Love you."

"Love you too," he said.

I closed my eyes after hanging up and took a deep breath. It was a long ride up to the twenty-first floor that seemed to go on and on. When the elevator finally settled and the doors opened, Julie was waiting for me off to the side in the corridor.

"How are you feeling?" I asked.

"Excited," she said. Then: "Nervous."

"That's normal," I said. "I'm nervous too. You ready to go?"

She nodded.

We walked toward the conference room where many of the board members were already gathered out front. I shook a few hands and made polite introductions between Julie and our directors, many of whom didn't know her. I explained that she'd be joining us for part of the meeting today.

A few minutes later, at 2 PM, I said bye to Julie and walked into the room with the directors. We took our seats around a long table to begin the meeting.

Richard spoke first.

"Welcome, and thank you for joining us today. As you know, this is a special board meeting—"

I interjected: "Richard, if you don't mind, could I say a few words to set the stage?"

Briefly, Richard's eyes narrowed as he scrutinized me, but ultimately he relented and said, "Go ahead."

"Thank you." I stood up. "A month ago, I walked into this boardroom expecting you to congratulate me. All of our metrics were up and to the right. Customer satisfaction, revenue, profitability, stock price. Every single metric. I thought I was doing everything right. But I was wrong."

I paused and glanced around the table. Ever so subtly, Paul seemed to be nodding, encouraging me to keep going.

"While I was focused on quarterly results, I missed that the basis of competition was shifting beneath our feet. AI is the new basis of competition. And I should have seen that sooner."

I looked directly at Richard.

"I owe you an apology. My job is to stay ahead of the curve, to see what's coming before you have to tell me. And I failed at that. But I also want to thank you. You did exactly what a board should do. You pushed me to focus on the right strategic priorities, even when it was uncomfortable for me. That's governance working as it should."

I took a deep breath, straightened my shoulders, and pressed on, confidence growing.

"The last thirty days haven't been easy. But I have a lot to share with you about what I've learned and where I envision us heading. To do that, I'd like to bring in someone who's been instrumental in developing our AI strategy."

Richard seemed lost for words, if only momentarily, and then said: "Well, I had prepared remarks, but I think I'll put those aside. They don't seem quite as relevant now."

He surveyed the room to get a sense of how people felt. Satisfied with his visual scan, he said, "We do appreciate the results you've been delivering, Sarah. The numbers are impressive—there's no doubt about that. And I respect your humility in taking on our challenge and acknowledging our governance role. That takes courage. Not every CEO would stand up and say what you just said. The floor is yours. Show us what you've been working on."

"All right, then," I said. "Let me bring in my co-presenter."

I walked to the door, opened it, smiled at Julie, who was standing off to the side, and beckoned her to come join us. As she entered, I introduced her.

"For those of you who haven't yet had the pleasure of meeting Julie Guthrie, allow me to introduce her. Julie is the president of our South Central Division, which, as you know, has been performing well. You may have also heard that her division is doing things with AI. But I doubt you know the lengths to which they have gone to not only use AI, but to use it effectively, responsibly, and in a way that creates meaningful and differentiated business value." Julie settled into her seat next to me and raised a gracious hand.

"It's a pleasure to be here with you all today," she said.

"Welcome," said Paul, with a polite nod.

Of the twelve board members, he was the only one to verbally welcome Julie. I heard the rustle of papers being adjusted. A muffled cough.

"Let's begin," I said. "First, I want to acknowledge that this company already has AI expertise, in the form of Julie and her team. We can build on that. Second, I want to point out there's a tendency to overpromise on AI—which is something I think companies like Apex are doing. Everything we're about to share with you, we've thought through in detail, and with the help of external experts who were able to help validate these ideas."

"Experts?" Diane, a usual skeptic, asked. "Did you hire a consulting firm?"

"We worked with an existing partner," I said.

"I trust them," Julie chimed in. "And I'd be happy to share the process by which we went about selecting them with you after this presentation."

That seemed to be enough to satisfy Diane, although she still seemed skeptical.

"FreightCo has a long history of excellent client service," I said. "For over thirty-five years, we've been a reliable logistics partner for the likes of major companies such as S&B. We want to use AI to improve their experience, while also expanding into new business lines. In addition to being a warehousing and trucking company, we're proposing expanding to become a broker as well as an orchestrator."

Richard raised his eyebrows. Paul looked on. Another board member began to interject about the scale of change, but Richard cut them off. "We'll ask questions at the end. Please, Sarah, go on."

"Thank you," I said. I went on to explain the vision to them in greater detail, and had Julie explain the finer points of the current routing solution and how it could be applied nationally. I emphasized that none of our competitors were currently configured to go to market in this way, and while Apex had a variety of announced AI investments, it wasn't altogether clear how they were doing things differently as a result. But this was different.

We presented the road map to the board, and yes, we'd have investments in other AI solutions, such as the intelligent contract review solution and co-generation tools. The board seemed particularly interested in these flashy examples of AI, but Julie steered them away from that and onto the investments called out in the bottom half of the road map: the enabling investments in things like talent and culture.

"It really cannot be overstated," she began, "how essential these things are to making AI work successfully."

She went on to talk about the culture of her division, and how she and August had worked to make it something intentionally distinct from the rest of the company.

There was some uncomfortable fidgeting and shifting in seats from some members of the board. I quickly reiterated: "We have a plan for AI governance and enablement to help institutionalize some of these . . . softer changes," I said. And then I added: "But make no mistake, they are essential changes and must be made. This plan will fail without them."

We concluded with a vision of FreightCo one year from now, where we'd established two new lines of business, retained over

95 percent of our customers from the year prior, increased revenue and profit, and, as a result, our stock price was the highest it had ever been.

When we finished, there was no applause. No shouts of congratulations. There was silence, and then there were questions.

"How, exactly, are we supposed to go from where we are now to doing all *that* in twelve months?" Diane asked.

"It does seem a bit . . . fanciful," Richard said.

"It's bold, that's for sure," I said. And then, looking at Paul, I said, "But we have a detailed plan. We have a template for how to do this at the division level. And we have a great partner." I checked my watch. It was 2:57 PM. Paul looked back at me, calm as could be.

"And I really think this plan is something that the market wants—both our current customers and truckers."

"That may very well be, but how much is it going to cost?" Richard asked. "Have you modeled the ROI?"

"We have," Julie said. "It'll take a little while to see the benefits. AI projects tend to require nonlinear gains. But once we hit twelve months, you should start to see the acceleration of progress, and money back in our pockets. We think we'd break even around that time, and things would take off from there."

"I'd like to review that plan," James, the venture capitalist, said.

"Of course," Julie said.

Richard cut in: "As I said before, we appreciate you putting together this plan, but—" He looked around the table, getting a sense of people's reactions. "I hope you can understand that we still have a lot of questions."

"I do," I said. "Please, ask away."

"Okay," he said. "I can see the appeal for truckers if we became a broker. But for our current clients, wouldn't they be reluctant to sacrifice control of their operations and pay a premium at the same time? How do you know this would work?"

Just then, at 3 PM, my phone rang.

"Excuse me," I said. "Do you mind if I take this? It's the CEO of S&B."

Richard gave me a look that asked, *What the hell are you doing?* But he didn't say no.

"Go ahead," said Paul.

I stood up and took the call. The room was absolutely still, and people stared at me intently. You could hear a pin drop.

"Yes—it is. Thank you for calling," I said. "Do you mind if I put you on speakerphone?" Then I addressed the room. "Is everyone okay if I put Phillip on speakerphone?"

Nobody said a word, so I did.

I briefly summarized the context for him: We were in a board meeting discussing our exciting plans for how we intended to use AI to serve our customers.

"Very good," he said. "We were deeply impressed with what you showed us. It was bold, and we liked that, and it was believable, which is a nonnegotiable for us, particularly when it comes to AI. We also have a lot of confidence in Julie."

"She's with us now," I said.

"Hi, Phillip," Julie said.

"Julie! Thanks again for the excellent presentation. We plan to stick with you. I'll have our folks get started on making it official later today."

"Fantastic. Thank you for letting us know," I said.

We hung up.

"So?" I asked.

Richard turned back around to get a sense of the feeling in the room. It didn't take him long.

"We're running short on time, but everyone appears very satisfied, myself included. Well done," Richard said. "Let's meet again next month and continue discussing the details of the plan."

"Although, before we adjourn," Paul spoke up, "I think we should include board AI governance as a topic in our next meeting. This strategy needs the right board oversight structure."

Richard agreed.

With a deep sense of relief, I cleared my throat and said, "One more thing: As many of you know, there's no way we would be here without Julie's AI leadership—she deserves the lion's share of the credit. Given that today went so well, I think I can go ahead and make this announcement—" I looked at Richard, who nodded.

As I looked around the room, I said, "I'd like to introduce you all to FreightCo's first Chief AI Officer. Do you want to say a few words?"

"Thank you," Julie said. Her voice was softer than during the presentation but steady. "I appreciate the recognition. And I must admit that I'm a little bit intimidated, to tell you the truth. But I'm excited to help lead FreightCo into this new age of AI, and to do it ethically and responsibly."

After a brief round of applause, folks stood up from their seats and shook our hands. There were slaps on the back, and a claim of "We always believed in you" as well as an omission of "I wasn't sure you were going to be able to pull it off." Michael asked for information on Guidebridge, and mentioned he'd like to float the idea of working with them to three of his other companies.

Paul came over to me and gave me a hug.

"Never in doubt," he said.

"It was for me," I said. "I doubted."

"Well, that's all part of this, isn't it? This big game."

"I suppose so."

Julie was accepting her plaudits on the other side of the room.

"Impressive, that one. She'll make a great CAIO," Paul said.

"She will," I said. "Although I'll need to figure out how to support her in the role, and who I'm going to get to backfill her position as head of her division."

"You will," he said. "You always do."

"I really wasn't so sure this time," I said. "I genuinely thought I was going to get fired because I was so unsure of how to get a handle on AI."

Paul paused for a moment, stepping back to consider what I'd said. "But you didn't."

"I know. But I thought I might."

"I'm sure there must have been real ups and downs for you this past month," Paul said. "You should take some much deserved time off."

I hesitated, rolling my pen back and forth in my palm, considering how to respond. He gave me that knowing look of his.

"There will be plenty of time to go fast later," Paul said. "Rest and recharge. Plan a vacation. Fill out a crossword puzzle with that thing." He gestured to my pen.

"The first thing I'm going to do is fill out Julie's promotion paperwork. And then I'll see where I'm at."

Paul smiled.

"But you're right—I do need some time to recharge and reflect," I added, and it began to dawn on me that I had, up until now, thought of the board presentation as the finish line. But it wasn't. It was a milestone, a moment to savor amidst the long and untidy process of finding our way.

Epilogue

Julie and I had done a few of these now, but I still got nervous.

"Three minutes until you're on," said a stagehand.

I turned to Julie. "We got this."

"We do."

And still, the feeling of going onstage, of the walkout itself, knowing there was only one direction you could go and that was forward, to face the audience—the one-way-ness of the action intimidated me.

"Water?" Mark offered, having returned with some.

Julie accepted, and thanked him.

"Remember," he said, trying to seem calm. "Deep breath, big smile. Do you want to do a power pose together? We could do that right now."

"Two minutes."

"Sure, why not?" I said. And we all stood, shoulders back, eyes closed, taking one deep breath, and exhaling, arms stretched out as if to embrace the world.

"All right, then," he said. "Looks like they're coming to get you. Good luck."

I thanked him, and couldn't help but think that he'd really taken his role of helping people across the company learn to work with AI and made it uniquely his own. The man who had once pretended to have all the answers had grown into the kind of leader AI requires, one whose vulnerability and humility allow him to not know all the answers, ask for help, and create the space for others to learn without fear of judgment.

I followed Julie out onstage. The lights were bright, and I squinted to try to make out David—there he was—sitting next to Paul in the second row. Ava was with them, as was Oliver, who had been right a year ago when he speculated that Apex was bluffing. A series of incidents, some high-profile, others mundane, had severely hampered their business.

Julie and I settled into large, plush chairs.

Our interviewer, a young man who seemed happy to see us despite the fact we'd not met before, welcomed us to the Keynote for AI Leadership.

After brief introductions, he asked the first question.

"Sarah, as many in the audience know, FreightCo has had a stellar year, with the launch of two new business lines, impressive growth, and a reputation that can't be matched. While I'm aware that you—and your Chief AI Officer, Julie—have been speaking and writing about your approach to AI, which you call 'Wayfinding,' many of us here today would love to hear firsthand just exactly how you managed to pull off this remarkable transformation. Can you tell us more about how you got to where you are today?"

"Sure," I said. "For me, it started with a boardroom ultimatum. But Julie was doing the hard work long before I was, and luckily, I was able to learn from her."

And then, as had become natural for us, I passed it to Julie to tell the story.

Appendix

Welcome to the *Wayfinding* appendix!

The original and proprietary methodologies, frameworks, and learnings in *Wayfinding* are hard-won from eight years of in-the-trenches experience of the team at KUNGFU.AI, an AI-native professional services firm developing leading-edge strategies and bespoke AI solutions on the cutting edge of the technology for organizations large and small across a wide range of industries.

While we hope that you have picked up the learnings we've tried to convey throughout our business fable, we also wanted to give you all of the information in one place in a chapter-by-chapter breakdown.

Introduction

- **Core Learnings:**
  - Success, as measured by traditional metrics, can blind you to fundamental changes in the basis of competition.
  - AI is fundamentally changing the basis of competition.

- **Detailed Explanation:**
    - When Sarah walks into the boardroom expecting congratulations for all of FreightCo's metrics being up and to the right, she's caught off guard by the board's ultimatum. This is not because traditional metrics are irrelevant. But, in this case, they're lagging indicators. Her board is well positioned to spot the disruptive threat posed by AI, and have chosen to vocalize their concern, even if their delivery is a bit stark.

Chapter 1

- **Core Learnings:**
    - Perfectionism stymies progress, and taking the first step to act likely means being able to admit that you don't know something.
    - Leaders struggle to let go of legacy mental models, and emotions are the "silent blockers."
    - Your frontline employees can be great sources of information—tap into them.
- **Detailed Explanation:**
    - Sarah's struggle in this chapter is meant to reflect something we see a lot with our clients: Executives naturally rely on mental models that made them successful, but AI requires updating their mental models, which they resist doing. After all, why do away with the set of beliefs and actions that made you successful, particularly when you feel threatened and need to respond to that threat? When Sarah hesitates to text

Mark and Julie because she doesn't have perfect questions, we're seeing her perfectionism in action. Her fear of "looking silly" stalls progress, albeit only for a short while thanks to Paul's encouragement. Paul's coffee shop intervention forces her to name these emotions and move past them. His advice to "start by talking to people who are already doing the work" before her inner voice convinces her to research endlessly is a direct challenge to her mental model. The fear, embarrassment, and dread she feels when considering embarking on her journey aren't character flaws. Instead, they're the natural response of someone whose entire career was built on having answers suddenly confronting a domain where nobody has all the answers. Booking the flight to Austin before she feels ready is a small gesture of change. The chapter title, "Starting Over," is, itself, an homage to Sarah having to rebuild her leadership identity.

Chapter 2

- **Core Learnings:**
  - Talking about AI and actually using AI are two totally different things—and in AI, where failures are inevitable, those unprepared to handle them will be exposed.
  - Substance beats ".ai" hype. Cosmetic moves like rebranding erode credibility and can alienate your customers and employees.
  - Traditional software processes tend to be insufficient to

properly train, deploy, and monitor AI solutions. Sticking rigidly to existing ways of working only to have them backfire underscores the need to adopt a learning mindset and not pretend to have all the answers.

- **Detailed Explanation:**
  - Mark Armstrong is an archetypal character who embodies common mistakes we see in the field, such as an insistence on having all the answers, a reluctance to admit mistakes or learn from others, a tendency to confuse doing things with actual progress, a tendency to prioritize marketing hype over real results, and a rigid focus on applying traditional methodologies to AI development. While Mark is ultimately redeemed, largely thanks to the care and attention afforded to him by Sarah (and his family and therapist), his behavior creates havoc along the way. He loses clients, alienates colleagues, and wastes time and money, to name a few consequences of his actions. Luckily, he has a CEO who chooses to lean into being vulnerable with him at the hotel bar and is thus able to finally break through to him and create space for him to be vulnerable in return, thereby initiating his redemption arc. When Mark returns a changed man, Sarah goes a step further than simply reintegrating him to the team and instead tests his new willingness to lean into being wrong by discussing his team's failed AI pilot, and then, as it dawns on her that Mark has many valuable skills, not least of which is the gift of relentless optimism and (at times self-) promotion, Sarah

gives Mark a new role working with Ross and August to help enable the rollout of AI across FreightCo.

Chapter 3

- **Core Learnings:**
  - Instead of asking broad questions about what AI can do, build the ability to succeed with AI by starting with specific business problems that already cost you money.
  - Human-machine teaming often works better than full automation because humans and AI have complementary skill sets.
- **Detailed Explanation:**
  - When Julie walks Sarah through the contract review system, she's demonstrating a principle of AI development that sounds obvious but that most companies miss: Pick a real problem you already have and try to solve that. Contract reviews were eating up her team's time (particularly on routine work, while complex matters piled up). That's a specific, measurable problem with clear costs. The solution isn't fancy—the AI flags sections with color codes and suggests replacement text while Anita makes final decisions. She still catches things the AI misses, like when she knows a client always requests specific warranty language for regulatory reasons. This is human–machine teaming in practice: The AI handles pattern matching while humans handle exceptions and context. The technical detail about RAG (retrieval-augmented generation) matters

here—Julie's team didn't just use AI "generically" but connected a language model to their specific context (their contracts) so it could reference actual company language. Julie mentions this almost casually, but it's an important distinction. When Julie admits they were "disappointed" with early results until they connected it to their contracts database, she's being honest about the iterative nature of AI development. The contrast with Mark is instructive: He's chasing "full automation" and losing clients, while Julie's "boring" approach of augmenting human work is actually delivering value.

Chapter 4

- **Core Learnings:**
  - Data readiness is unglamorous but foundational as having data isn't the same as having usable data.
  - Plant the AI "data tree" today.
  - Reversible decisions with human oversight are ideal starting points for AI experiments.
  - Success requires clearing calendars and dedicating resources, not just buying technology.
  - AI that works can threaten other teams, making internal politics a hidden obstacle.
  - Creating psychological safety—where people can admit not knowing and openly share failures—is a prerequisite for success with AI.
  - In tight AI talent markets, culture is your best form of sustained competitive advantage.

- **Detailed Explanation:**
  - Julie Guthrie represents a type of leader who succeeds with AI transformation, embodying important qualities such as intellectual humility despite deep expertise, patience to build foundations before flashy wins, and the courage to protect her team from corporate pressures. Her admission that she "literally lay down on the floor" overwhelmed by AI's implications is something we've heard before from clients as well as an indication that she is willing to be vulnerable with Sarah (and is very unlike Mark, who needs to appear to be in control). Furthermore, her response to the feeling of being overwhelmed is admirable: methodically attending webinars, interviewing her team, bringing in consultants. She is someone who treats not-knowing as a starting point rather than a weakness. Julie openly admits spending months just cleaning data because "AI systems are only as good as the data they're based on." She also planted the data tree early, collecting information "even before we hired Ross" because she recognized data as an asset even before knowing exactly how she'd use it.
  - The contract review system captures Julie's pragmatic philosophy: It's reversible ("we could always revert to manual review"), it helps rather than replaces valuable employees, and it solves a real problem without theatrical promises. When she reveals trying twice to share approaches with Mark only to have him talk "at her" rather than "with her," we see someone who has

learned when not to waste energy. Her description of maintaining an "island" against the "ocean" of corporate culture reveals the exhausting reality of innovation in risk-averse organizations. Additionally, she notes that others in FreightCo have felt threatened by her work. It is not uncommon for AI—particularly AI that works—to make leaders and employees alike feel threatened, be it because it renders prior investments obsolete, threatens jobs, or reflects well on another team.

  - Ultimately, it should be noted that Julie is not perfect. The qualities (humility, patience, protectiveness) that have brought her success when Sarah meets her at the beginning of the story have also made her less visible in the organization and less willing to collaborate with those outside of her division. Throughout the story, she grows into a character more willing to delegate, champion successes, and, ultimately, take on a new role as Chief AI Officer.

Chapter 5

- **Core Learnings:**
  - AI investments can be "invisible" to outsiders and may let competitors pull ahead unnoticed.
  - There's a strong market incentive to be perceived as doing AI, which can lead leaders to make AI announcements before they have actual AI capabilities.
  - Data flywheels reward early movers.

- **Detailed Explanation:**
  - During the dot-com bubble, a lot of progress was visible. For instance, it was possible to visit rivals' websites and gauge their e-commerce maturity against yours. But with AI, this often isn't the case. Much of the pivotal work happens out of sight, which is to say, critical AI investments—data pipelines, model training, etc.—can be invisible to outsiders. Thus, when Apex announces their "AI-Native Logistics Platform," it blindsides Sarah. She assumes Apex has been quietly building these capabilities, though we later learn they were largely bluffing.
  - This reveals a different dynamic: Because markets react to visible announcements while real AI work remains hidden, leaders feel pressure to announce capabilities before actually building them. There is also an irony to Apex's announcement of AI "success" coming so soon after Sarah's trip to Austin: FreightCo has its own invisible AI success that even the CEO didn't fully know about.
  - The data flywheel makes invisibility particularly dangerous. Companies that get ahead accumulate better data, build better models, attract more customers, and generate more data—creating a compounding advantage. Once behind, catching up becomes exponentially harder. When Paul mentions engineers getting "offers that are hard to refuse," it reveals something unsettling: For all their bluster, Apex does have very good competitive intelligence, enough to know to poach members of Julie's team!

Chapter 6

- **Core Learnings:**
  - There's a difference between recognizing a threat and knowing how to respond to it.
  - Crises can accelerate timelines and force decisions before anyone is ready.
- **Detailed Explanation:**
  - The emergency board meeting is meant to dramatize a phenomenon we often see in organizations: the desire to act without knowing exactly what to do. In part, this can be driven by fear—of being left behind, of not understanding something—but it can also be driven by conviction and can be quite positive (after all, as we later learn, having a strong AI mandate is essential to a successful AI strategy). Ultimately, it is Paul in this chapter who is meant to serve as the role model: He alone seems to recognize that learning must precede action, he is calm amidst the chaos of the meetings, and he has done significant prework to buy Sarah more time.

Chapter 7

- **Core Learnings:**
  - Building support requires vulnerability and asking for help from a wide variety of people.
- **Detailed Explanation:**
  - This chapter is meant to convey that different people

experience AI differently, and that comprehensive learning requires soliciting diverse viewpoints. For instance, between Sarah and her children, we see a stark generational divide: Jake casually mentions using AI for homework "all the time" while defending it as a legitimate tool when "everyone else is" using it. To him, AI is just part of the landscape. But to Sarah, it's an existential threat requiring a herculean sprint to save her job!

- David's response is meant to model what supportive partnership looks like during a professional crisis. His observation in the introduction that Sarah is "treading water" rather than drowning reframes her struggle as temporary rather than terminal, and he continues to support her in this chapter by affirming how aggressive her new thirty-day timeline is. When Sarah reaches out to Julie, Mark, and her CEO friends (Dante, Noah, Maya), she's building a learning network that spans divisions, companies, and industries. Each brings a different perspective that is meant to serve as a composite of leadership experiences with AI. Furthermore, the decision to invite Mark to Austin despite his failures shows Sarah beginning to understand that AI transformation requires bringing together diverse perspectives, even if you don't agree with them. Ultimately, navigating AI isn't a solo journey. It requires admitting vulnerability to family, building coalitions with unlikely allies, and recognizing that even teenagers might understand aspects of this technology better than CEOs.

Chapter 8

- **Core Learnings:**
  - AI differs from traditional software in fundamental ways, and these differences cascade through many aspects of how you build and manage it.
- **Detailed Explanation:**
  - When Oliver and Ava of Guidebridge.ai present their table comparing traditional software and AI, they're addressing a misconception that derails most AI initiatives: treating AI like traditional IT. The framework they share covers nine dimensions.

How AI Is Different from Traditional Software

Aspect	Traditional Software	AI/ML
User Experience	Deterministic	Probabilistic
Adaptability	Requires discrete updates	Able to be updated using new data
Performance Basis	Performance based on code	Performance based on code and usage context at time of inference
Decision-Making Basis	Logic-based (more transparent and interpretable)	Data-driven (can be a black box)
Development Approach	Linear	Iterative
Project Management Approach	Waterfall; traditional scrum; etc.	Wayfinding (hybrid agile)
ROI	Focused on efficiency gains, sometimes new revenue	Focused on efficiency gains, new revenue, and a valuation premium for tech leadership
People	Hire experts; talent impact is typically linear	Hire learners; talent impact is exponential
Culture	Plan and execute culture; psychological safety is desirable	Learn and adapt culture; psychological safety is needed to encourage experimentation

- **User Experience:** Traditional software tends to follow clear rules and creates a deterministic user experience—the same input always produces the same output. AI models tend not to follow programmed logic but instead predict what response is most likely based on patterns learned from training data. As a result, similar prompts can result in wildly different outputs, which, to a user, feels probabilistic, like throwing dice.
- **Adaptability:** Traditional software requires code changes and new releases to improve. But AI can be updated by feeding it new data or fine-tuning it without having to rearchitect the system. This makes AI more flexible but also means it can drift from its original behavior.
- **Performance:** Traditional software performance depends on how well the code is written. AI performance depends on both code *and* the context at the time you use it—the prompt, the data it's referencing, even the specific examples you provide. This is why Julie's contract system needed connection to their specific database to be useful.
- **Decision-Making:** Traditional software follows explicit logic that you can often trace. AI, because it does not follow intelligible logic, can be a "black box," meaning it might give you an answer without being able to explain why. This creates challenges for regulated industries and situations requiring accountability.

- **Development and Project Management:** Traditional software development tends to aim for linear progress using methodologies like waterfall or traditional scrum. AI development is inherently iterative with nonlinear progress—hence "Wayfinding," our flavor of agile optimized for AI projects (see Appendix Entry, Chapter 12). Like other agile methodologies (scrum, kanban, XP), Wayfinding emphasizes iteration and rapid feedback. But unlike traditional scrum that pushes for immediate development, Wayfinding acknowledges AI's unique needs, such as spending more up-front time for data exploration and hypothesis testing (what agile practitioners call "Sprint Zero").
- **ROI:** Traditional software ROI focuses on efficiency gains and sometimes new revenue. AI can deliver both as well as a valuation bump for being seen as a tech leader—which is why Apex's stock jumped on their announcement alone.
- **People and Culture:** Traditional software rewards expertise and planning. AI rewards people who admit they don't know and can adapt. In this way, psychological safety is paramount (you literally cannot succeed with AI if people can't admit what they don't know). Furthermore, top performers in AI can have an exponential rather than a linear impact, making hiring and then setting those people up for success all the more crucial.
- This chapter's core message is that you can't succeed with AI using traditional software methodologies.

Chapter 9

- **Core Learnings:**
  - AI transformation requires three interlocking components—Strategy, Operations, and Engineering—working as a flywheel where each reinforces the others.
  - Bad consulting experiences can poison an entire organization against AI, making partner selection critical.
  - Organizations tend to fail when they over-index on one component (like strategy) while neglecting others (like operations).
- **Detailed Explanation:**
  - The chapter opens with Noah's cautionary tale: His board hired consultants who "slapped ChatGPT onto their existing services and raised their rates," burning $100K on a useless road map. Now his board thinks AI is "snake oil" and won't approve another dime. Sadly, this is an experience we've heard about happening in organizations far too often. The hype around AI has given rise to many less-than-qualified "partners." Working with such a "partner" can lead to a single bad experience that embitters an entire organization against working with external AI partners. This is particularly bad as working with a trustworthy, qualified AI partner has been shown in studies to dramatically increase the likelihood of success with AI.
  - We intended to further underscore the thematic juxtaposition of AI charlatanism by having Mark—a snake oil salesman himself—enter this chapter just

as Sarah is talking to a fellow CEO who has been burned by snake oil salesmen.

- Back in the conference room, Oliver and Ava (trustworthy consultants) present the AI transformation framework we tend to use with clients, which depicts three overlapping circles that form a flywheel.
  - **Strategy** requires leadership alignment around AI initiatives and a strong mandate to pursue AI, even if initially unfocused. Without executive sponsorship, every failed experiment becomes ammunition for skeptics. In this case, the board's ultimatum, while harsh, at least provides the mandate Sarah needs.

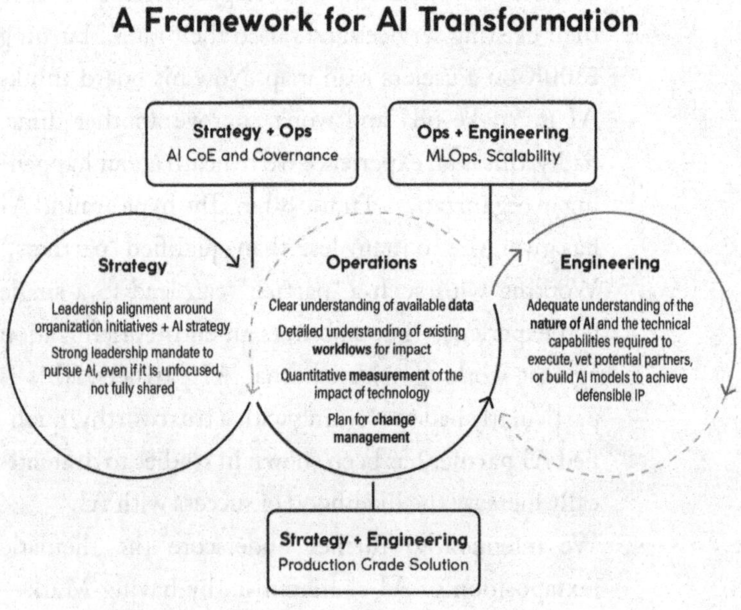

  - **Operations** establish foundations to build, deploy, and sustain AI advantage. This means understanding your data assets, mapping current workflows to know what to augment, measuring impact quantitatively, and planning for change management. As Ava warns, "This is where most AI initiatives actually fail"—companies discover they don't have the right data, or it's siloed, or they don't understand their own workflows well enough to improve them. Julie spent months on this unglamorous work, including two months just on change management for the contract system.
  - **Engineering** involves having technical capabilities to execute AI initiatives, either internally or through carefully vetted partners. Julie found the right partner in Guidebridge after vetting multiple firms. We learn about how she did this in the following chapter.
- In the AI transformation framework, the magic happens at the intersections. Strategy + Operations is, in effect, an AI Center of Excellence. Operations + Engineering affords you MLOps and scalability. Strategy + Engineering can produce a production-grade solution that has the potential to generate real business value, but without operations, deploying it and measuring impact is difficult.
- Thus, it is through the activation of all three disciplines that organizations can achieve AI transformation.

- Mark's resistance throughout ("Sounds nice on paper," "We don't have time for all that hand-holding") represents the impatience that tends to kill AI initiatives. When he suggests rebranding to "FreightCo.ai" while dismissing the foundational work, he embodies the preference for AI theater over transformation. Oliver's reference to Amara's Law—that we tend to overestimate technology's short-term impact and underestimate its long-term impact—is meant to rebuff Mark and level-set expectations.

Chapter 10

- **Core Learnings:**
  - Because there is so much noise in the vendor space, selecting the right AI partner requires methodical evaluation.
  - AI is a toolbox, not a single tool. It spans much more than just generative models, including things like predictive analytics, computer vision, optimization, and more.
  - The best AI partners prioritize building your internal capabilities rather than creating dependency.
- **Detailed Explanation:**
  - Over dinner, Julie reveals how she learned to identify real AI expertise. Her framework focuses on two categories: maximizing opportunities and minimizing risks. The questions she asked are the questions we

generally advise our clients to ask when selecting a partner.

- Oliver's origin story mirrors Stephen's experience cofounding KUNGFU.AI. As Oliver explains: "Guidebridge came out of my pattern recognition from the internet era." Stephen lived through the same pattern—watching the commercial internet from 1996 onward change the basis of competition and topple established firms. By the 2010s, he saw identical patterns emerging with AI.

Characteristics of an Ideal AI Professional Services Partner

They help maximize outcomes because they:	They help minimize risks because they:
· Are focused on creating significant and differentiated value. · Have deep capabilities across the entire AI toolbox. · Help their clients transform into AI-centric organizations. · Meet clients where they are. · Take care to build a deep understanding of their clients' businesses.	· Are truly AI focused. · Can help with strategy, engineering, and organizational transformation. · Are investing in keeping up with the technology to stay on the leading edge. · Understand the need to be responsible with AI.

- Stephen's key insight was that, unlike the internet era, where software engineers could relatively easily pick up web development, cutting-edge AI requires rare skill sets fundamentally different from traditional programming. He wanted to build differentiated AI models based on proprietary data—creating enduring competitive advantages for clients, not just temporary efficiency gains. This vision required a new kind of firm: a "mash-up" combining rare AI engineering capabilities with traditional strategy consulting, all focused on delivering differentiated business value.

- When Stephen cofounded KUNGFU.AI in 2017—years before ChatGPT made headlines—it wasn't obvious AI would become the transformative force it is today. But he believed it represented the next major wave of technological change, and here we are (let it be noted that I, Ben, am grateful he took that leap and cofounded the company!).
- Furthermore, when Ava mentions "AI isn't a tool—it's a toolbox," she's articulating something we often underscore with clients. Most executives hear "AI" and think "ChatGPT," which is to say, "GenAI," but AI includes image classification, recommendation systems, anomaly detection, and many other techniques, as shown in the handout Oliver gives Sarah.

The Full AI Toolbox | Part 1

Analytical AI

Vision & Documents
Image: Object Detection; Classification, Change Detection, Segmentation, Facial Recognition; Pose Estimation

Video: Action Recognition, Video Understanding; Tracking

Documents: OCR, Layout/Table/Field Extraction, Handwriting Recognition, VQA

3D: Depth Estimation, Structure from Motion; 3D Object Detection; Volumetric Segmentation

Language & Code
Text Understanding: Classification, NER, Part-of-Speech Tagging, Topic Modeling, Sentiment/Intent Modeling

Retrieval: Embedding Models, Vector Similarity, Approximate Nearest Neighbor, Hybrid Lexical & Semantic Search, Re-Ranking

Speech & Audio
Automatic Speech Recognition; Speaker Recognition; Emotion Recognition; Diarization

Audio Event Detection; Source Separation; Sound Localization

TTS, Voice Conversion; Voice Cloning

Tabular & Time-Series
Classification, Regression, Forecasting, Anomaly Detection; Imputation; A/B Testing

Recommender Systems

Decision Science: Survival Analysis, Uplift Analysis, Causal Inference

Graphs
Node/Edge/Graph Classification

Knowledge Graph Construction

Community Detection; Graph Anomaly Detection

Generative AI

Language and Multi-Modal: LLMs for Text Generation, Vision-Language Models (VLMs)

Code Generation & Review

Retrieval-Augmented Generation (RAG), Context Engineering

Agentic Systems (functions, APIs/MCPs, browsers, CLI, planners, verifiers)

Image/Video: Diffusion Models for Image and Video Generation (T2I, I2I, T2V, I2V, etc.), Image Captioning, GANs, VAE, Navigational World Models (action-conditioned video generation)

Audio: Audio Generation & Editing (speech & music), Speech-to-Speech

3D: NeRF, 3D Gaussian Splatting (3DGS)

Math: Theorem Proving

Biology: Protein Design, Structure and Interaction Prediction

Chemistry: Molecule and Material Design

Physics: Fast Physics Surrogates for PDE Solvers

The Full AI Toolbox | Part 2

Physical AI

Digital Twins

SLAM, Navigation

Vision-Language-Action Models (VLA)

SIM-to-Real

General Algorithms and Best Practices

Responsible AI
Alignment: RLFH, Constitutional AI, Etc.

Red-Teaming: Jailbreaks, Prompt Injection, Bias/Safety Stress Tests

Transparency: Model Card, Data Sheet, Feature Attribution, Mechanistic Interpretability

Privacy: Federated Learning, Differential Privacy

Model Architectures
ConvNet, LSTMs

Transformers, State-Space Models (SSMs)

Gradient Boosted Trees

Kernel Methods

Generalized Linear Models

Training Algorithms
Pretraining, Supervised Fine-Tuning (SFT)/Instruction Tuning, RL with Human Feedback (RLHF), RL with Verifiers

LoRA, Model Distillation

Self-Supervised/Contrastive Learning

Inference Optimization
Model Quantization

Speculative Decoding, KV Caching

Unsupervised & Semi-Supervised Learning
Clustering

Dimensionality Reduction (PCA/UMAP/t-SNE)

Active Learning

Model Deployment/Productionalization
MLOps: Experiment Tracking, Shadow Deployment, A/B Testing, Deployment, Performance Monitoring

Orchestration: Model Context Protocol

Evaluation and Benchmarking: Model Evaluation, Human-in-the-Loop

Containerizing: APIs, Etc.

Chapter 11

- **Core Learnings:**
  - The characteristics required for AI success (humility, vulnerability, curiosity, admitting ignorance) are also useful dispositions to have in one's life.
- **Detailed Explanation:**
  - Mark's breakdown at the hotel bar reveals the connection between his professional failures (see Appendix Entry, Chapter 2) and personal crisis. The same inability to admit "I don't know" that led him to fake AI success is destroying his marriage—his wife says he's "distant" and "disconnected" while he doesn't understand what she means. When Sarah shares her own marriage struggles and how therapy helped, she's modeling the vulnerability that Mark desperately needs to learn. She recognizes that Mark can't succeed professionally until he addresses what's happening personally. Mark's willingness to finally ask for help marks the beginning of his transformation from AI charlatan to someone who might actually contribute meaningfully to FreightCo's AI journey.

Chapter 12

- **Core Learnings:**
  - AI development follows an S-curve—slow discovery, rapid progress once aligned, slow deployment—not

the linear progress associated with traditional software development.
  - AI models optimize to "win at all costs," typically without understanding context, thus exploiting any pattern in data (including ones you never intended).
  - Wayfinding is a flavor of agile designed for AI's unique needs where more time is spent on up-front exploration and iterative problem discovery.
- **Detailed Explanation**
  - Wayfinding is our adapted agile methodology specifically designed to address the unique challenges of AI development. The framework is predicated on the observation that AI projects fail when managed like traditional software projects because AI development is inherently different. Ross's routing story is meant to illustrate this. His team's AI found "efficient" routes that were mathematically sound but nonetheless brittle in the real world (they'd fail at the first road closure). This happened because AI models optimize ruthlessly to "win" based on their training data, without understanding real-world context. They'll exploit *any* pattern, including biases or correlations you never intended.
  - The Wayfinding framework consists of a series of steps that form an iterative cycle. In the Discovery Phase, we have Alignment (defining the actual business problem), Exploratory Data Analysis or

EDA (determining if you have the data to solve it), and Solution Validation (testing if your approach works). After that phase, we have the Technical PoC Phase (building a working prototype), which includes defining requirements for the solution ("Requirements Defined") and conducting model training ("Training"). Following that, we have the Business PoC (proving the solution can scale and still produce the outcomes you desire), and finally, Productionalization (actually putting it into production). But unlike waterfall, you're expected to iterate back to earlier phases if you discover that your assumptions are wrong (which is typical). And a significant percentage of time is spent in the first four phases, defining the business problem to be solved, exploring the data needed to solve it, testing models, and then, finally, proceeding to making a Technical PoC. That first funnel is shown below, highlighted in gray:

Wayfinding

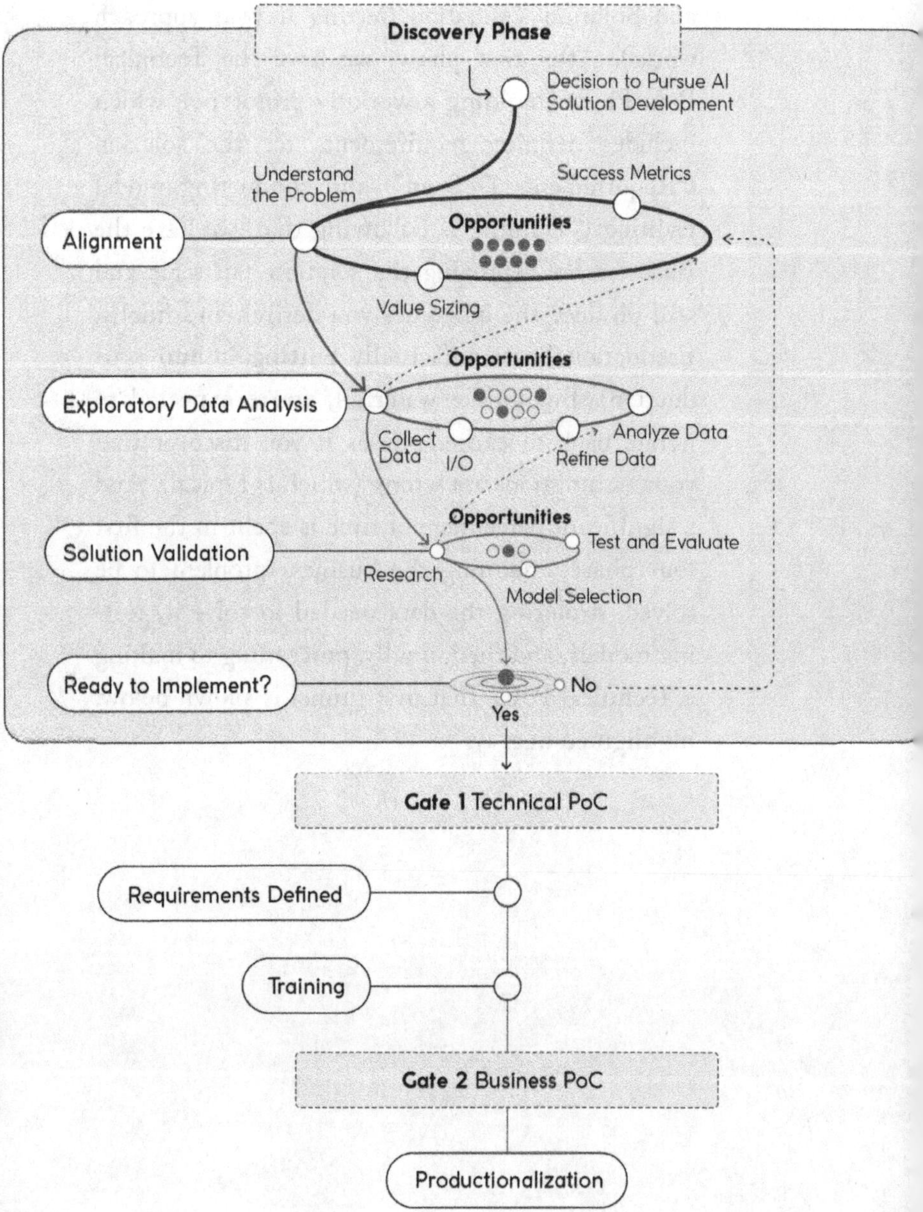

Discovery Phase
Decision to Pursue AI Solution Development
Understand the Problem
Success Metrics
Alignment
Opportunities
Value Sizing
Opportunities
Exploratory Data Analysis
Collect Data
I/O
Refine Data
Analyze Data
Opportunities
Solution Validation
Research
Test and Evaluate
Model Selection
Ready to Implement?
No
Yes
Gate 1 Technical PoC
Requirements Defined
Training
Gate 2 Business PoC
Productionalization

- Our PoC-to-production success rate using this methodology (90 percent) versus the industry's dismal average (around 10 percent) is due, in part, to accepting these realities up front.
- Another aspect that makes Wayfinding different from traditional scrum is that it acknowledges and actively embraces the S-curve as the ideal model of development progress when it comes to AI projects (shown below).

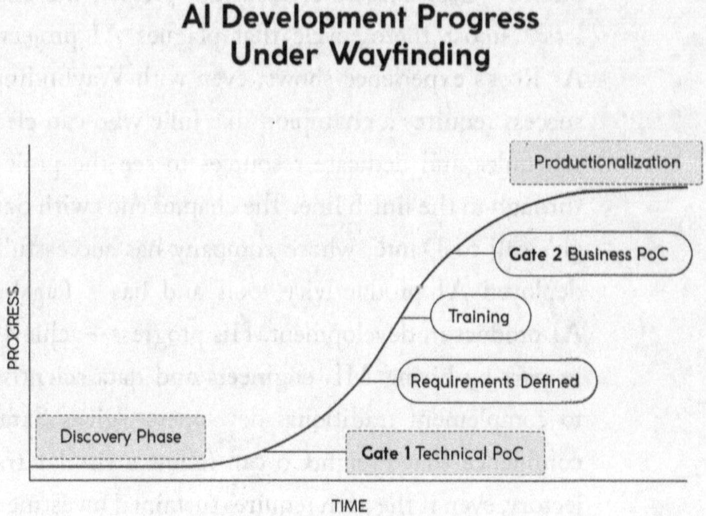

- Where traditional scrum expects steady sprint-by-sprint advancement, we find that AI projects tend to progress slowly during initial Discovery while you figure out what you're actually solving and whether you have data for it, accelerate rapidly once you understand the problem and the model starts learning, and then slow again during the "last mile" of Productionalization when you're handling edge cases and ensuring reliability. To use an analogy: If Waterfall is planning every detail of a road trip before leaving, and pure Agile is jumping in the car unprepared, then Wayfinding is setting clear objectives while accepting that you'll need to recalibrate constantly based on what you learn.

- The framework also emphasizes "gates" between phases—clear success criteria that prevent the endless "almost there" cycle that plagues AI projects. As Ross's experience shows, even with Wayfinding, success requires a champion like Julie who can clear schedules and dedicate resources to see the project through to the finish line. The chapter ends with Sarah's call to Dante, whose company has successfully deployed AI productivity tools and has a flagship AI product in development. His progress—achieved in part by hiring ML engineers and data scientists to complement traditional developers—gives Sarah confidence that FreightCo can follow a similar trajectory, even if the path requires sustained investment and new talent.

Chapter 13

- **Core Learnings:**
  - Top AI talent has exponential rather than linear impact.

    - Secret use of AI creates an "elephant in the room" that requires proactive acknowledgment and guidance.
    - Vulnerability from leadership (admitting you don't know) helps create the conditions necessary for AI transformation.
    - Managers face disproportionate disruption. They must learn new tools, help others adopt them, and some face threats to their status.
    - First AI hires are critical: A players hire A players; B players hire C players. But you can't simply hire your way to AI success. You must help existing employees evolve.
- **Detailed Explanation:**
    - August's revelation that they just lost Andy Bertrand to Apex—their second ML engineer poached this year—immediately establishes the stakes of attracting and retaining talent. However, when August says "you can't build a retention strategy based on keeping everyone," she's being pragmatic about a harsh reality: In tight AI markets, you'll lose people. The key is creating an environment where most choose to stay despite higher offers elsewhere.
    - The chapter's pivotal moment comes when Julie describes standing up at an all-hands to admit she didn't understand AI but would figure it out with her team. This vulnerability—anathema, perhaps, to traditional executive behavior—contributes to the early conditions for success in Julie's division. Julie's transparency steered the energy driving secret ChatGPT

use (something we see all the time) into a visible, shared exploration of what AI can do.

- Furthermore, her division's weekly pulse surveys are helpful not only in simply quantifying the employee experience, but also in doing it on an extremely granular level, which is important because AI accelerates everything, including changes in employee sentiment.
- When August mentions giving ML engineers Friday afternoons for upskilling, she's acknowledging that in a field moving this fast, investing in continuous learning can be a great retention strategy. This is also an homage to a practice we have at KUNGFU.AI—Lab Day—where we dedicate every Friday first and foremost to learning and development such that we can stay at the cutting edge of the quickly developing field of AI.
- The discussion of hiring reveals another crucial insight: First AI hires set the tone. As August notes, "A players hire A players; B players hire C players." Getting your first hires wrong compounds mediocrity.
- Another of the insights in this chapter concerns managers. They're experiencing "disproportionate change" because, on top of having to learn new tools, they also have to help others do so, manage teams differently, and (for some) deal with threats to their status. Many managers have built careers on having the right answers. But now they must model not-knowing. This

is a hard thing to do. So hard, in fact, that August admits to not having "cracked it yet" when it comes to supporting managers through this transition.

Chapter 14

- **Core Learnings:**
  - Stepping away from a crisis to synthesize learnings can be essential for breakthrough thinking.
- **Detailed Explanation:**
  - Sarah watching Emily's track meet while trying to process everything she's learned is meant to serve as a brief vignette about how insights coalesce. At first, in response to David's gentle questions, Sarah begins to spiral, thinking about Julie potentially leaving for Apex (thus showing what happens when we try to problem-solve while exhausted). But when Sarah walks away and reflects on her daughter's warm-up, and the value of taking space clicks for her, she creates the mental space where disparate pieces can begin to connect. Ultimately, when she returns to Austin in the ensuing chapter, she is more grounded and her level of knowledge about AI and her ability to manage the situation is improved. This chapter is meant to remind us that AI transformation is a marathon, not a sprint (so to speak) and that leaders who don't step back to synthesize and recharge often burn out before the finish line.

Chapter 15

- **Core Learnings:**
  - Teams can only go so far without deep trust. Achieving breakthrough performance requires leaders who are willing to be vulnerable with each other.
- **Detailed Explanation:**
  - The breakfast conversation is meant to reveal how a culture permeated by a lack of trust has limited FreightCo's potential. Sarah probes Julie's career ambitions, worried she might leave for Apex, while Julie deflects with "I see myself helping you prepare for that board meeting," a nonanswer that maintains distance. When Sarah asks directly, "What keeps you here?" Julie responds with confusion. We are seeing two people test each other's intentions. Sarah needs to know if she can count on Julie; Julie needs to know if Sarah will finally support her properly.
  - Although their working relationship has thus far been productive, Sarah and Julie's mutual wariness has nonetheless kept them from fully achieving what they're capable of together. Julie developed a complete twenty-four-month AI road map—presumably much of what the board wants—but hasn't shared it because she's worried Sarah would take it without understanding the effort behind it. Her admission that talking about their work "hasn't exactly been rewarding" reveals her disposition to protect her team rather than seek organizational support.

- The turning point comes through vulnerability. When Sarah finally apologizes—"I didn't know what I was missing. And that's my mistake"—she's taking the first risk. Julie's response is measured but revealing: She appreciates having a female CEO but notes they haven't worked closely together, acknowledging both the potential and the distance. When Julie finally shows the road map with its detailed value-at-stake analysis, she's reciprocating that vulnerability.

Chapter 16

- **Core Learnings:**
  - AI strategy must be tied to overall business strategy—it's not a separate initiative but an enabler of how you compete—and is typically comprised of a vision, a readiness assessment, and a road map.
  - Well-guided AI investments have the opportunity to do at least one of the following: increase revenue, increase profit, or increase differentiation. Hitting all three (the "Trifecta of AI ROI") tends to drive higher valuations as well.
  - Success requires both direct AI investments (tools and solutions) and enabling investments (data, talent, culture)—most organizations focus on the former while neglecting the latter.
- **Detailed Explanation:**
  - This chapter walks through multiple frameworks that together form a comprehensive AI strategy.

- The team starts by establishing that AI strategy has three components: a vision aligned with business strategy, an assessment of readiness, and a road map of initiatives.

The Components of an AI Strategy

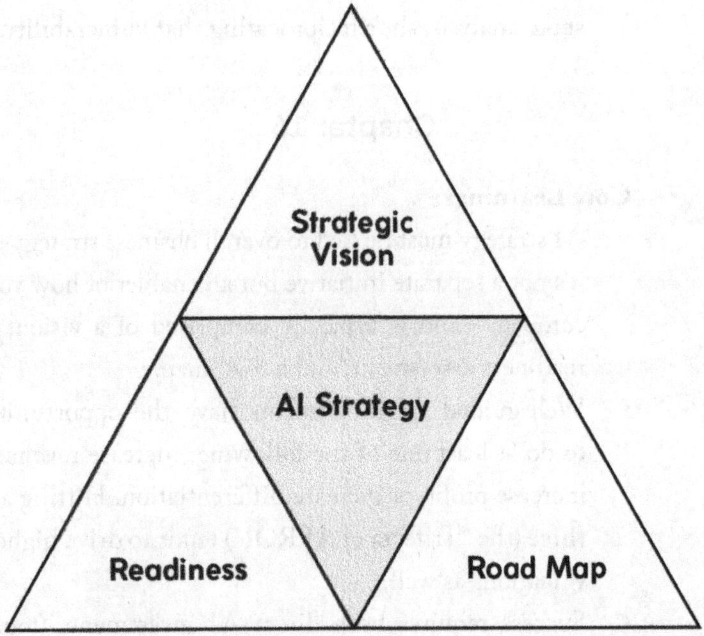

- Julie demonstrates this with her division's approach, showing how business strategy ("How we want to compete") must connect directly to AI strategy ("How we can use AI to compete").

AI Strategy Should Support Business Strategy

- The team then develops core principles: AI must be pervasive across the organization, investments go beyond technology to include data and culture, and initiatives should build long-term capability rather than just completing projects. Their readiness assessment reveals FreightCo's typical profile: high executive buy-in (thanks to the ultimatum), medium investment capacity, low AI talent except in Julie's division, and unknown data readiness.

The Main AI Value Drivers

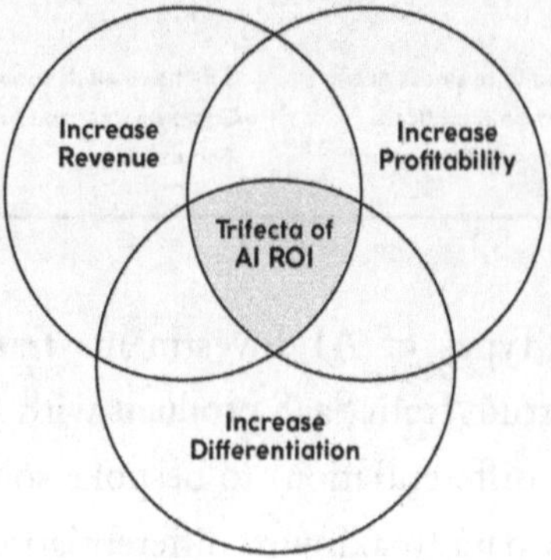

- The AI ROI framework—the "Trifecta"—shows how AI creates value beyond cost-cutting. Revenue grows through better products and services reaching more customers, profit through efficiency gains or improved products and services that can command a price premium, but it is differentiation that creates a lasting advantage and higher valuations (explaining Apex's stock jump).

Types of AI Investments

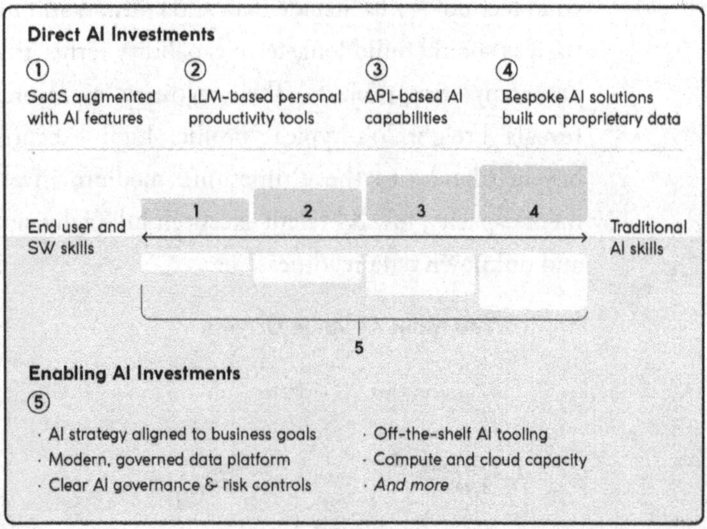

- The Types of AI Investments framework covers a spectrum from SaaS products with AI features (minimal differentiation) to bespoke solutions on proprietary data (maximum differentiation). Critically, the framework shows that enabling investments—data platforms, governance, talent, culture, etc.—determine whether direct investments succeed.

Chapter 17

- **Core Learnings:**
  - A change of place (or pace) gives leaders room to process complex information and return with renewed clarity.
  - Organizations are failing with AI in opposite ways—some paralyzed by risk aversion (Maya's company), others creating chaos through recklessness (Mark's division).
  - Leaders who have failed at AI can become valuable assets if they're willing to be transparent about their mistakes.
- **Detailed Explanation:**
  - Sarah's evening run along Lady Bird Lake is meant to give both her and the reader a break from the dense chapters before and after this one. It is also a true-to-life experience that I, Ben, have had many times in Austin (running post–work commitment). For Sarah, running gives her the space to allow her mind to wander, for her energy levels to return, and, crucially, for her to not have to be anyone to anyone else. Simply put, she can just be. We don't view this as avoidance of responsibility. Rather, it is the creation of space to process complex events. When Sarah's roles (CEO, mother, wife) fall away and she becomes just another

runner, she's able to synthesize the day's learnings without the weight of expectations.

- Upon returning to the hotel, her two phone calls provide important contrast. Maya's board is taking a "wait and see approach" with AI—all governance, no enablement—the opposite of Mark's chaos. When Sarah explains her situation, she notices she doesn't feel as tense as usual, in part a consequence of the run she just went on as well as a sign that the strategy is solidifying in her mind.
- Mark's call continues his transformation arc. He's lost two more clients but now takes full responsibility and wants to learn from Julie. Sarah's decision to give him another chance isn't just an act of kindness—it's also strategic. Mark represents FreightCo's broader culture, and if he can transform, others can too. The chapter shows that leadership during transformation isn't just about driving forward. It's about knowing when to step back, let your mind process things, and return with the clarity to make difficult decisions.

Chapter 18

- **Core Learnings:**
  - Effective AI governance balances risk management with enablement.
  - The best AI strategies don't engage in a race to the bottom on head count reduction and cost-cutting (even if they incorporate those elements), but instead

reimagine business models and create new value streams with AI.
  - Sustainable AI advantage tends to come from leveraging your unique assets (data, relationships, domain expertise) in ways competitors can't replicate.
- **Detailed Explanation:**
  - Mark's return opens the chapter in dramatic fashion. He last left this room theatrically, but now, he stands still, makes eye contact, and has the self-awareness to suggest adding "don't burn credibility chasing hype" to their AI strategy principles. His apology to Julie and admission about his failures sets the stage for using those failures as learning opportunities.

An Organization-Wide Framework for AI Governance

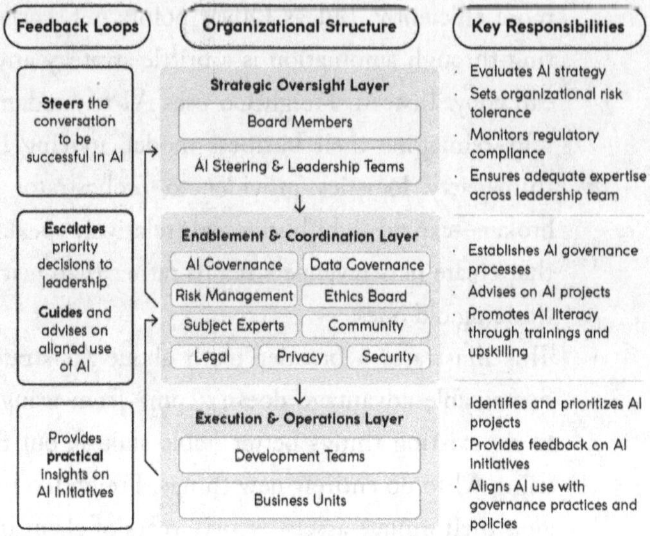

- The ensuing governance discussion reveals how to avoid both extremes—Mark's previous chaos and Maya's paralysis. The three-layer model they design creates structure without imposing stifling bureaucracy: strategic oversight (board and leadership setting risk tolerance), enablement and coordination (centers of excellence guiding adoption), and execution (teams implementing with feedback loops). As Ava explains, governance should be like traffic signals—telling you when it's safe to go fast and when to slow down, freeing people to focus on driving rather than agonizing over every decision.
- The strategic breakthrough demonstrates a crucial principle: compete on your own terms, not your competitors'. Apex promises "99.9 percent optimization"—essentially doing the same things more efficiently. But as Oliver points out, cost-cutting through automation is a brittle strategy anyone can copy. Instead, FreightCo uses AI to fundamentally reimagine their business model, moving from third-party logistics provider to orchestrator and broker—capital-light businesses (relatively speaking) that create new revenue streams rather than just optimizing old ones.
- This illustrates a broader truth about AI strategy: Sustainable advantage doesn't come from using AI to do existing things better (table stakes), but from using AI to do entirely new things. FreightCo leverages their unique assets—thirty years of client trust,

proprietary route data, deep operational knowledge—in ways Apex can't replicate. When Julie suggests predicting which deliveries need "five-star service," she's showing how AI enables business model innovation, not just operational efficiency. Mark's assignment to lead workforce education proves that reformed skeptics, who understand the pitfalls firsthand, can become powerful change agents, thereby completing his transformation arc.

Chapter 19

- **Core Learnings:**
  - Leading with vulnerability and taking responsibility creates space for authentic transformation.
  - Demonstrating early wins with real customers validates strategy more than any slide deck.
  - Promoting internal champions to formal AI leadership roles signals organizational commitment.
- **Detailed Explanation:**
  - Sarah's opening move—apologizing to the board for missing the AI shift while thanking them for pushing her—transforms the room's dynamics. By taking responsibility rather than defending her record, she demonstrates the vulnerability that AI leadership requires. This isn't weakness; it's the strength to admit what you don't know and commit to learning.
  - The presentation itself follows a crucial sequence: First, acknowledge existing internal expertise (Julie's

team), then show realistic assessment of competitors (Apex is likely overpromising), and finally present a differentiated vision that builds on unique strengths. But the masterstroke is Julie's orchestration of the S&B call (alluded to at the end of chapter eighteen). When their biggest client confirms they're staying because FreightCo's AI strategy is "bold" and "believable," it vindicates Sarah and Julie's approach and leaves little room for debate.

- The board's approval without even taking a vote shows how compelling real results can be. But Sarah's final move—announcing Julie as Chief AI Officer—is equally important. This isn't just a well-deserved promotion for Julie. It's a clear move to begin to institutionalize a different way of working. By elevating someone who built AI success from the ground up rather than hiring externally, Sarah signals that FreightCo values substance over credentials. The chapter demonstrates that successful AI strategies shouldn't only live on paper. They should be validated through real-world interaction and must be implemented by competent, trustworthy people.

Epilogue

- **Core Learnings:**
    - Having success with AI is not a one-and-done thing. It is measured by the organization's ability to continuously learn and adapt.

- Wayfinding is both a technical methodology for managing AI development and a metaphor for building the skills to be a great leader in the age of AI.

- **Detailed Explanation:**
  - One year later, Sarah and Julie are speaking at the keynote of an AI Leadership conference—a reversal from the opening where Sarah knew nothing about AI. FreightCo has launched two new business lines and avoided the fate of companies like Apex (whose overpromises led to "a series of incidents").
  - Mark's transformation into the person leading AI education and adoption efforts completes his redemption arc. The man who once pretended to have all the answers now embodies the vulnerability and humility that "creates space for others to learn without fear of judgment"—a complete reversal from the defensive executive who used to feign competence rather than admit ignorance. Well done, Mark!
  - The word *Wayfinding* carries dual meaning throughout the story. In the literal sense, it's a framework for developing AI solutions. But as a metaphor, it captures the essence of leadership in the AI age: You're not following a predetermined map but rather finding your way through uncertain terrain, recalibrating based on what you learn. Wayfinding requires humility, patience, teamwork, and perseverance. It is also, as you have likely noticed, the title of this book.
  - Sarah's passing of the microphone to Julie in the final paragraph of the book is meant to encapsulate the

journey they've been on together. Ultimately, this is a story about three leaders—Sarah, Julie, and Mark, each with their own strengths as well as weaknesses to surmount—who found their way by learning to trust one another.

Acknowledgments

While the cover lists two coauthors, and it's true that this book is the result of a true collaboration, I want to acknowledge that Ben Szuhaj wrote the entire thing. So how did this all work?

It started in the spring of 2024, when someone at KUNGFU.AI—where I'm cofounder and CEO—suggested that I write a book. I quickly pooh-poohed the idea. But when someone, a few months later, brought the idea up again, but this time suggested I write a business fable in the style of Patrick Lencioni, a favorite of mine and a giant of the genre, I again quickly rejected the idea, but it did stick in my head.

And there the idea stayed, with me barely thinking about it at all for months, until I woke up on my last day of vacation with my wonderful wife in Yellowstone, positively bursting with the full vision for the book. I spent the whole day traveling home, pounding out six pages of an outline and synopsis on my laptop.

And that's where the idea stayed for another few months because I'm self-aware enough to know that I'm only a halfway decent writer and someone with a short enough attention span to know that a project as big as writing a book is very hard for me.

But then I learned that Ben, one of the newer members of our AI strategy team, had studied creative writing in college, loves to write, and aspired to one day publish a book. And so we decided to collaborate on the project, with me providing the overarching vision and a lot of the content, and Ben filling in the gaps and weaving it all together with great characters and plot twists along the way.

With that as background, I want to give a huge shout-out and thank-you to you, Ben, for being a terrific partner on this project and for single-handedly bringing it to fruition!

Separately, I want to thank my incredible cofounders—Steve Meier, Ron Green, and ZZ Si—and the whole team at KUNGFU.AI, for all they do every day to deliver significant and differentiated value to our clients and help make the company such a great place to work!

—Stephen Straus

Together, we'd like to acknowledge that the team at KUNGFU.AI developed the methodologies we share in this book, based on eight years of delivering leading-edge AI strategies and bespoke solutions. We'd also like to thank our clients who we have had the privilege of serving as we help build significant and differentiated value for them.

This book has also benefited significantly from the people who took the time to talk to us about their experiences leading AI and/or read early drafts and gave us valuable feedback, which we greatly appreciate. More specifically, we want to thank clients

Heffy Provost, CEO of AGDATA; Susan Winslow, CEO of Macmillan Learning; and Jason Heilig, CTO of Triumph; company advisors Steve Rohleder and Ralph Hasson; teammates Reed Coke, Michael Wharton, Christian Barnard, Edward Cates, Ryman Stringer, Benjamin Herndon, Daniel Bruce, Emma Pirchalski, Nathan Mandi, and the many others at KUNGFU.AI who contributed their time and expertise; as well as friends of the company Jim Triandiflou and John Arnott.

We also want to express our deep gratitude to our publisher, Matt Holt, for believing in us and this project, as well as the whole team at Matt Holt Books and BenBella Books, including our editors, Lydia Choi and Katie Dickman; cover designer, Jason Arias; production editor, Isabelle Rubio; publishing associate, Ariel Jewett; marketer, Mallory Hyde; vendor content manager, Alicia Kania; and sales manager, Susan Welte. And a big thank-you to Eric Siegel for the original introduction to Matt!

Finally, we deeply appreciate the patience, support, constructive criticism, and love of our friends and families. In particular, we'd like to thank our partners, Tina Straus and Sophie Bennett, without whom this book wouldn't have happened. And for Ben, his parents, Patricia Keleman and Timothy Szuhaj, as well as Rachel Ruina and Andy Bennett.

—Stephen Straus & Ben Szuhaj

Our Use of AI in the Writing of This Book

Given this is a book on AI, we wanted to share how we used generative AI models in our writing and what we learned in the process. First, we want to emphasize that we wrote this book, not AI. But we did use generative AI models throughout to support our efforts. In some respects, our creative process mirrored the book's central point: that deriving real value from AI requires an open mind, experimentation, and a propensity to reimagine ways of working.

We used Google's Gemini to transcribe video we recorded while playacting certain scenes, and then used Anthropic's Claude to pull out specific compelling details that we then used to add depth and, hopefully, realism to those scenes. We also used Claude in a brainstorm capacity and as a sounding board for ideas, as well as to suggest line editing and ways to refine what we wrote. Claude also categorized what we'd written, which helped us manage the spreadsheet we had created with the topics we wanted to weave into the story.

At the same time, we found AI to be seriously unhelpful for actually drafting compelling, original fiction prose. The AI-generated prose tended to be overly recursive and melodramatic—in a sense, almost a parody of what we were trying to create.

All in all, we found numerous ways that generative AI was able to serve as an able assistant in our writing.

About the Authors

Stephen Straus is the CEO of KUNGFU.AI, a boutique AI management consulting and engineering firm in Austin, Texas, that he cofounded in 2017. Stephen brings a wealth of knowledge and stories from over 150 AI strategy and engineering engagements delivering significant and differentiated value to clients in over thirty industries. Stephen is a serial entrepreneur and former venture capitalist and is a graduate of Colgate University and Harvard Business School. He and his wife are recent empty nesters with three great daughters and a son.

Ben Szuhaj is an AI strategist and writer passionate about helping organizations navigate AI transformation. At KUNGFU.AI, Ben works with leaders to develop and implement practical AI strategies that deliver tangible business value. As a writer and speaker, Ben excels at making complex technological concepts accessible and engaging. Ben holds a BA in English with a minor in Human-Centered Design from Dartmouth College, where he won the Mecklin Prize for creative nonfiction or journalism, and is a competitive long-distance runner.

To learn more about Stephen, Ben, and KUNGFU.AI, please visit www.kungfu.ai.